T5-ASC-386

New technology, skills and management

In recent years, industry has had to respond to rapid developments in new technology to survive in an increasingly competitive environment. Fast, effective adoption depends on successful training programmes. While in the long-term new technology can play an active part in the development of a continuous innovation process and in strategic decision-making.

New Technology, Skills and Management analyses the various ways training programmes can contribute to a company's success as it adopts new technology. Based on studies of the engineering and micro-electronic industries, it covers areas such as the relationship between skills shortage and unemployment showing how the lack of training opportunities can have a detrimental effect on the company performance and efficiency of its workforce. It argues for the importance of training not just in the workplace but in the educational system in general, concluding that there needs to be much greater investment in training opportunities.

This book will be of interest to students involved in human resource management, industrial sociology, applied economics, occupational psychology and industrial relations as well as practising managers and policy-makers.

Adrian Campbell is Lecturer in Management, University of Birmingham. He previously taught and researched at Aston University, London Business School and Henley Management College where he completed his PhD in management studies. **Professor Malcolm Warner** is a Fellow of Wolfson College, and a member of the Judge Institute of Management Studies, Cambridge University. He is the author, co-author and editor of twenty books and over 100 articles in learned journals.

New technology, skills and management

Human resources in the market economy

Adrian Campbell and
Malcolm Warner

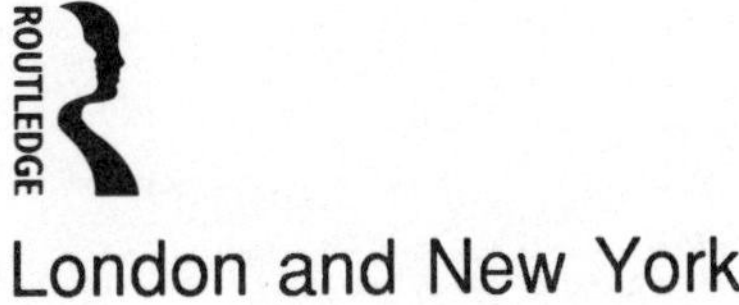

London and New York

First published 1992
by Routledge
11 New Fetter Lane, London EC4P 4EE

Simultaneously published in the USA and Canada
by Routledge
a division of Routledge, Chapman and Hall, Inc.
29 West 35th Street, New York, NY 10001

Typeset by LaserScript Limited, Mitcham, Surrey
Printed and Bound in Great Britain by
Hartnolls Limited, Bodmin, Cornwall.

British Library Cataloguing in Publication Data

A catalogue reference for this book is available from the British Library.

ISBN 0–415–05555–5

Library of Congress Cataloging in Publication Data
has been applied for

ISBN 0–415–05555–5

Contents

Figures and tables

FIGURES

TABLES

Preface

This monograph contains the findings of a field study carried out by the authors which took as its focus the implications of microelectronics in product applications for company human resources strategies, skill-profiles and requirements, with a parallel study conducted in West Germany by a team from the International Institute of Management (IIM) in Berlin West, the results of which are reported elsewhere (Campbell *et al.* 1989a). This represents a change of emphasis from a previous Anglo-German project (Sorge *et al.* 1983).

We would like to thank all those who assisted in the project in different ways – not least the large number of managers, trade union officials, experts and others, who agreed to be interviewed and who, unfortunately must remain anonymous. Their help and co-operation was very much appreciated. Among the friends and colleagues we should thank are John Barber, William Brown, Wendy Currie, Colin Gill, Alex Lord, Riccardo Peccei, Sheila Rothwell, Wolfgang Streeck and Jonathan Zeitlin, as well as the IIM researchers, Werner Beuschel, Sabine Gensior, and most notably Arndt Sorge. We would also like to thank those we talked to in the Department of Trade and Industry, the Training Agency, all of whom were very helpful. Next, many thanks to Jill Ford, the project secretary and Pauline Johnson who typed the manuscript. We should also like to thank those academic institutions which generously gave their support in the preparation of the final manuscript, particularly the University of Birmingham, the Judge Institute of Management Studies, (University of Cambridge) and Wolfson College, Cambridge.

Finally we would like to gratefully acknowledge the assistance and support of the Anglo-German Foundation for the Study of Industrial Society which funded the project, and Hans Wiener, its former Deputy-Director, whose encouragement throughout this and previous research endeavours has been unstinting.

Abstract

This study sets out to find how national culture and institutional tradition influenced the ways in which companies and their managements in the engineering industry in the United Kingdom responded to changes in product-technology and markets in the 1980s, in particular the way in which their skill requirements were fulfilled.

The results of the British study reveal that the introduction of microelectronics into engineering products coincided with a crisis regarding mass-production markets which led to a greater degree of competition and an increase in customer expectations; in addition, the trend towards customisation drew both on these events as well as on the increased automation that has occurred in design, through the use of Computer-Aided Design (CAD) and Computer-Aided Engineering (CAE) to enable 'repeats' and modifications. These developments in turn facilitated a greater emphasis on design rather than manufacturing capability, with design and production becoming increasingly separate activities in many cases.

As with all the findings of the study, counter-trends were found to be in force – the new developments, particularly in small-batch or unit production, could mean a merging of the two functions. While the small-batch production entails a greater emphasis on skills than the tradition of British manufacturing is geared to, developments in product-technology and in design-automation make customised production more akin to mass production in order to reduce costs. The availability of key skills, particularly in the electronics area, appears to have become a major constraint on companies.

In general, the human resources strategies of companies placed too much emphasis on graduate recruitment rather than the training of technicians to fill the same roles, which in the latter case could often be to their longer-term advantage. Graduate retraining, in the light of technological developments, also had to be expanded. In these respects popular conceptions of the skills-shortage have proved to be simplistic. It was also found that the increased overall complexity of technical systems, as opposed to the structure of their component parts which has in many cases been simplified through microelectrics applications, has led to a need – as yet relatively unfulfilled for more broadly trained hybrid engineers with a high level of product awareness, and appropriate management strategies to cope with human resources development.

Acknowledgements

We would like to thank the following publishers and journals for permission to cite from material previously published:

Gower Press
Human Systems Management
Industrial Relations Journal
Journal of General Management
Kogan Page
New Technology, Work & Employment
University of Wales Business & Economics Review
J. Wiley and Sons

Abbreviations

AEU	Amalgamated Engineering Union
ASICs	application-specific integrated circuits
ATE	automatic testing equipment
AUEW	Amalgamated Union of Engineering Workers
CAD	Computer-Aided Design
CAE	Computer-Aided Engineering
CAM	Computer-Aided Manufacturing
CAP	Computer-Aided Planning
CIM	Computer-Integrated Manufacturing
CNC	Computer Numerical Control
CSS	Computer Special Systems
CSSE	Customer Services Support Engineering
DNC	Direct Numerical Control
DTI	Department of Trade and Industry
EETPU	Electrical, Electronic, Telecommunication and Plumbing Union
EITB	Engineering Industry Training Board
ESSA	Electrical Supervisory Staffs Association
FMS	Flexible Manufacturing Systems
HRM	human resources management
IIM	International Institute of Management
IPG	International Products Group
IOSG	Information Office Systems Group
IT	information technology
LSI	large-scale integration
MOD	Ministry of Defence
MSC	Manpower Services Commission
MSF	Manufacturing, Science and Finance Union
NAC	Networks and Communications

NC	numerical control
NCVQ	National Council for Vocational Qualifications
PCB	printed circuit board
TECs	Training and Enterprise Councils
TVEI	Technical, Vocational and Educational Initiative
VET	Vocational Education and Training
YTS	Youth Training Scheme

Chapter 1

The shifting ground of the debate

This book reports the findings of a study of the relationship between new technology (in the form of product, as opposed to process applications of microelectronics) and human resources strategies (HRM) of skill provision. It led the authors into several different debates, some of which have receded or moved on, whilst others are currently reaching their zenith in both political and academic terms.

In the first category we may place questions such as whether new technology creates or destroys jobs, whether or not it de-skills workers, and the nature of its effect on the division of labour in industrial society. These debates, which had a high profile when the study was initiated in the mid-1980s have to some extent yielded their position as the political consensus in favour of technological innovation in manufacturing has consolidated itself, winning over not a few sceptics in the process. However, given their continuing underlying relevance – particularly in relation to the debate over 'flexible specialisation' which sprang up in the second half of the last decade, and their influence on the political and theoretical climate in which this study was undertaken, a review of the literature relating to them has been included in Chapter 2.

THE PLACE OF TRAINING IN THE MARKET ECONOMY

In contrast to some of the debates referred to above, the issue of training has seen an ever-increasing saliency in the late 1980s and early 1990s. During this period criticism by opposition parties of government industrial policy has increasingly focused on the issue of training and skills as a key area of government weakness. There has been a growing consensus, from trade unions to employers' associations and beyond, that matters such as training and research and development cannot be left to market mechanisms alone. Indeed, as we argue in our closing chapter, market mechanisms may actually work against training through a version of the 'prisoner's dilemma' effect.

Human resources policies based on recruitment from the external market mean poaching, and poaching acts as a deterrent to training not a stimulant (contrary to what would be expected from a traditional market forces viewpoint). There is even now a temptation to regard the problem of poaching as a non-issue, or a reflection of carelessness on the part of the losing firm – 'If a firm trains and is losing people, it has likely got its training right and everything else wrong', as the manager quoted by Hendry (1991) puts it.

This view correctly draws attention to the non-pecuniary reasons for employee turnover and the consequent need for employers to approach human resource policy in a broader and more imaginative way, to take account of a more educated workforce. However, it ignores the extent to which the system as a whole (that is, the linkage between the external recruitment market and the internal labour markets of firms) is failing to operate effectively.

The 'systemic' failure across whole sectors of the British economy is recognised even by commentators such as Maynard (1988) who is broadly positive regarding the government's record on the economy:

> the 'free-rider' problem remains, namely that firms not spending on training can still derive benefit from the spending of firms which do and there could be advantage in making a levy on all firms to finance industrial training programmes
>
> (Maynard 1988: 168)

This statement implies a profound criticism of the government policy from 1979 onwards, in particular that regarding the training boards, including the abolition of many of them (see Rainbird 1990).

There are perhaps two fundamental problems with the strategy currently adopted nationally. First, there has been a degree of bureaucratic and ideological confusion over what successive training initiatives are intended to achieve, leading to some incoherence (Senker 1990). Some initiatives – the Youth Training Scheme (YTS) and Employment Training (ET) are examples – have appeared to have the primary aim of reducing youth unemployment rather than improving economic performance. Others, such as the Technical, Vocational and Educational Initiative (TVEI) were loosely based on German models, but insufficiently well thought-through to have a significant effect (Senker 1990). Initiatives such as curriculum reform and National Council for Vocational Qualifications (NCVQ), city technology colleges and the changes in primary, secondary and higher education seemed to pull in two contradictory directions, with some aspects emphasising the aim of wider access, while others entrenched further the traditional British principle of 'early extraction of elites' (Sorge and Warner 1986).

DERREGULATION VERSUS A COHERENT MARKET

The second problem has been the concern to infuse training policy with a rather literal view of the way free markets operate. Marsden and Ryan (1989) have, for example, described the way in which this interpretation of the free market prevented the development of the YTS into the regulated occupational scheme it might have become. The government's goal appeared to have been the establishment of low youth wages and a free hand to employers to carry out the training as they saw fit. During the same period they allowed apprenticeships to wither almost into insignificance. As it happened, industry still demanded occupational (market-wide) rather than internal (company-specific) qualifications. The result was that the demand for apprenticeship-trained workers far exceeded supply and YTS failed to provide a substitute.

As Marsden and Ryan conclude:

> A central problem with public policy towards initial training in Britain lies in its neglect of the institutional context which is required to make it viable. The belief that deregulation will mean the creation of free markets and that free markets can solve training problems has been associated with the neglect of institutions, including internal and occupational structures in the labour market. . . . The result is the resurgence of skill shortages, both quantitative and qualitative.
>
> (1989: 17–18)

The point is taken up by Lindley (1991), who seeks consolidation of these occupational rather than internal labour markets. To do this, Vocational Education and Training (VET) policy would need to 'bring into clearer focus the roles to be assumed on the part of the individual, employer, VET supplier and government, fostering the individual's capacity to pursue an independent career' (1991: 225). For it to work 'the state should gradually move towards providing a minimum commitment of VET funding to each individual' (1991: 226), this minimum being spent at the individual's discretion.

This trend, rather than mere deregulation, would provide the basis for a genuine market. It would also increase access to education and training while maintaining transparency of qualifications and occupational mobility.

With deregulation, on the other hand, skill shortages lead to greater and greater mobility for that minority having appropriate occupational qualifications. A recent survey, cited by Stevens and Walsh (1991) found this to be more than 10 per cent. Meanwhile those without qualifications are forced into increasing passivity in the labour market, thereby exacerbating those same skill shortages – the same survey found as many as 40 per cent of firms reporting that their ability to carry out their business had been affected by shortages of information technology professionals.

The effects of this vicious circle of under-training were clear in our own study, and they form part of the 'low skills equilibrium' that Finegold and Soskice have seen as operating in Britain – 'a self-reinforcing network of societal and state institutions which interact to stifle demand for improvements in skill levels' (Finegold and Soskice 1988: 22).

Systems involving individual choice within a clear (usually tripartite) institutional and legal structure of VET have been applied with a generally high degree of success by many of Britain's industrial competitors, notably France, Germany and Sweden, but also by less obvious candidates such as Hong Kong (Keep 1989a). Curiously, information about these systems appears to have had little resonance in UK government circles, which remain by and large wedded to US models, with overwhelming emphasis on local employer-led initiatives such as Training and Enterprise Councils (TECs) (Keep 1989a).

Senker (1988, 1989) implies that there is a fundamental fallacy in government thinking here – namely, that an effective market mechanism should not be seen as synonymous with whatever private sector employers happen to choose to do. It is after all, he argues, the failure of many of these very same employers to train that contributed to the problem of skill shortages in the first place. Employers do not always respond to market signals in the way that a notional 'rational actor' would have done. Indeed as we have implied above, the short-term interests of employers have often discouraged them to train, even though there is little doubt that their long-term survival depends on it.

INTERNATIONAL COMPETITION AND EDUCATION

It is not our intention to imply that little has been achieved in recent years in the area of VET. Our own findings testify to the degree to which industry has been adjusting its attitudes and HRM strategies regarding training. Equally, in terms of skills supply, the percentage of people in the labour market with formal qualifications has increased markedly. Between 1979 and 1989, the percentage of people of working age with higher qualifications increased from 11.2 per cent to 13.6 per cent, while that for 'lower skills' qualifications increased from 20.6 per cent to 29.9 per cent. Intermediate skills performed less well, increasing from 19.4 per cent to 24.2 per cent (Stevens and Walsh 1991). The smaller increase in intermediate skills is important, since it partly invalidates the improvement in higher skills performance – the fewer the intermediate qualifications, the more likely it is that those with higher qualifications will be used less efficiently and therefore be in shorter supply – a tendency referred to by a number of our case study respondents. Stevens and Walsh argue that neglect of formal training has produced an 'intermediate skills gap' in relation to competitors such as Germany, where 90 per cent of school leavers are apprenticed to a *Meister* in one of 450 recognised trades (see *Guardian*, 9 June 1990).

None the less, the improvements, particularly at the lower end of the scale, meant that in 1988 only 19.5 per cent of 20–24 year olds had no qualification, as opposed to 37.3 per cent in older age groups (Stevens and Walsh 1991: 30). However, the basic pattern in Britain remains that of an above-average elite, and a below-average remainder. Pearson *et al.* (1990) underline this lopsided performance – a slightly above-average performance on graduate output (at 14 per cent) combined with one of the lowest rates for educational participation overall. In Britain, only 40 per cent of 18 year-olds are in either full or part-time education, compared to 80 per cent in Germany and 70 per cent in both Belgium and the Netherlands (Pearson *et al.* 1990).

SKILL POLARISATION IN BRITAIN

Higher education participation is likely (with some difficulty on the way) to reach the level of 25 per cent in coming years. Such an achievement would ameliorate only part of the problem. In contrast to its European neighbours, Britain now possesses a skills underclass of serious proportions, which belies its reasonable performance at higher education level. In the commercial and clerical area, for example, Britain has 5 per cent of employees qualified above A-level, the same figure as France. At intermediate level, however, the French figure is 49 per cent, against a mere 18 per cent for Britain.

Ormerod and Salama (1990) demonstrate that the polarisation of incomes in Britain is closely shadowed by a similar polarisation of qualifications – in other words, the less-skilled earn less. This fact is not surprising. What is less widely appreciated in Britain, where notions of a 'low-wage economy' and 'core and periphery labour markets' have enjoyed some currency in recent years, is the effect of new technology on these groups. Whereas those possessing basic and intermediate skills may suffer periods of cyclical unemployment, they are less likely to be made permanently unemployed through technology and structural factors, since the breadth of their skills enables them to adapt (age and regional factors notwithstanding). The unskilled on the other hand, whose work is more tied to routines, are likely to be swept out of all but the most menial tasks by increasingly cheap information technology. As Ormerod and Salama (1990) conclude: 'Unless literacy, numeracy and ability to communicate are augmented dramatically in the workforce, the underlying structural forces in the economy dictate that the underclass will grow significantly in the 1990s.'

DE-INDUSTRIALISATION OR FLEXIBLE MANUFACTURING

There is an irony here in that the de-industrialised information society, with its enforced leisure, has long found far more adherents in the UK and US than in continental Europe. De-industrialisation and permanent mass unemployment do not now seem to be an inevitable result of new technology (see Chapter 2), but can result from it in certain circumstances. Hirst (1989) has noted how some UK and US commentators have seen moves towards de-industrialisation as somehow leading the way in a technologically-determined process of development. The apparent unwillingness of other countries to follow this path suggests that the de-industrialisation concerned may be more a simple matter of failing to compete in manufacturing and failing to provide a VET system within which individuals can adapt sufficiently to technological change. Earlier in the last decade, it had been thought that a neo-Taylorist division of labour might operate between the First and Third Worlds, with the First World carrying out the 'information work'. This view did not have to wait long for its refutation; newly industrialised countries such as South Korea and Singapore followed the Japanese example and invested in higher and higher levels of qualification to take them into markets previously closed to them.

The view cited above was mistaken not only in its reading of international trends, but also in its belief that such an informational division of labour would be desirable or even practicable in a period of rapid change in technologies and markets:

> For all its 'post-industrial' trendiness, the view that tries to separate information and execution, R&D and production, depends on an old-fashioned view of manufacturing. It conceives of manufacturing as the physical execution of given tasks, and the model for this is the assembly line.
>
> (Hirst 1989: 160)

Contrary to predictions that information technology would centralise decision-making and bureaucratise skills further, the trend appears to have been the opposite. The complexity of technology and the need for rapid responsiveness to market changes may demand clearer strategic direction from the headquarters of an organisation, but the co-operation between functions needed to fulfil these demands necessitates decentralisation of operations.

DE-DIFFERENTIATION

Such a shift from centralised control is not predetermined of course – as Clegg (1990) has pointed out, flexible technologies will be used differently in modernist (mechanistic) and post-modernist (organic) organisations. Here may lie a clue as to why investment in process technologies such as CAD/CAM has not always had the same success in the US and the UK as it has in Japan or the 'third Italy'. However much the established organisational cultures of many 'Fordist' firms might resist the trend, or attempt to turn it in another direction, there does appear to be a global shift in the direction of self-supervision, flexibility and market discipline as a substitute for routine control. Clegg's (1990) intepretation of this trend emphasises the role of Japan in spreading a 'post-modern' approach to organisation, one which conforms neither to Eastern patrimonial or Western bureaucratic traditions.

The change appears to operate at a fundamental level – Lash (1989), for example has defined it in terms of 'de-differentiation'; what is occurring is a reversal of the long-established trend towards differentiation and formalisation of organisational structures and procedures.

DE-DIFFERENTIATION AND SKILLS

Campbell and Warner (1989b) described how this shift towards greater market-responsiveness required closer co-operation between function and greater 'hybridisation' of skills. A similar point is made by Worthington (1989) who comments on the extent to which design and production are seen as separate in Britain, with the vast majority of qualified engineers located in design rather than production. This result may be linked to Armstrong's (1987) criticism of the neglect of production in management and engineering syllabuses in Britain throughout much of this century.

There needs to be greater overlap, not simply between design and production engineering but also between engineering and commercial functions. The lack of communication and mutual comprehension between these functions has long undermined the performance of British industry. As Gunn (1987) has argued, competitive advantage now requires not only computer-integrated manufacturing, but computer-integrated marketing, with the boundaries between engineering, production and marketing becoming increasingly blurred.

This blurring of boundaries should not be seen simply in terms of co-operation between different groups of managers and specialists, it requires a more solid base of workers educated at least to intermediate level. The influential study of the furniture industry in the UK and Germany by Steedman and Wagner (1987) emphasised the advantages of broadly qualified labour not only in terms of productivity but also in terms of more flexible product strategy, developing the arguments of Sorge et al. (1983) further.

CURRENT TRENDS

The outlook for VET in Britain is becoming increasingly confused just as the consensus in favour of radical action is reaching into parts previously thought of as naturally opposed to it (such as the Institute of Economic Affairs – see *Financial Times*, 15 January 1990).

The sense of national emergency has apparently steeled the TECs (previously expected to be relatively conservative bodies) into a more active role, particularly when their resources were cut by 20 per cent (*Financial Times*, 3 December, 1990). Ironically, although TECs were apparently intended to circumvent the need for measures such as direct training levies, their complaints over lack of resources have if anything significantly strengthened the case for just such a levy (see *The Independent*, 3 December 1990, Guardian, 4 November 1991).

As support for the idea of a levy has grown, so has scepticism as to how the TECs and NCVQ initiatives will be able to cure the underlying weakness of the British labour market. According to Rose (1990b), these weaknesses are self-perpetuating – if there are too few trained workers already who will carry out the training? His concern is that this bottleneck will lead to an increase in the number of nominally qualified but fundamentally unskilled workers.

In the midst of the debate, the consequences of past neglect of training and investment have again hit the British economy which at the time of writing is once more entering the 'bottoming out' phase of a severe recession. It has led once again to companies cutting back on their training budgets, despite general acceptance that this will further weaken long-term performance (Smith 1990).

Ironically, the extent to which the current recession has hit retailing and finance along with manufacturing may to some extent have a positive benefit for skills in manufacturing. When our study was carried out, the expansion of information technology in retailing and financial services was exacerbating the difficulties manufacturing companies faced in terms of attracting and retaining recruits with skills in this area. On top of the direct loss to the higher paying service sectors, the manufacturing recession of recent years has in any case deflected students away from subjects like engineering and computing – the number of students listing computing as their first preference actually declined between 1981 and 1988 by 8 per cent, while the computing industry grew by 20 per cent world-wide during the same period (Keegan 1990).

It may be that the retrenchment and loss of glamour in retailing and financial services may make manufacturing once again an attractive option, although this depends on how far manufacturing companies really have altered their management philosophies to take account of higher expectation and educational levels among the workforce. To this confused picture has been added the ending of the Cold War. Just as British industry has been criticised for concentrating its talent in design engineering rather than production engineering and manufacturing, so it has also been criticised for the degree to which engineering talent is heavily concentrated in the military rather than the civil sector (Blackburn and Sharpe 1988).

The decline in the armaments market may lead to a freeing of specialists to work in consumer areas. However, this possibility has very likely arisen too late. Such a shift, had it occurred decades earlier, might have saved more of the British car industry or consumer electronics. As it is, much of the capacity in such sectors has disappeared, leaving the now-vulnerable sectors of defence and aerospace as the only major sectors where Britain has consistently performed well in world market terms. Whether British industry can effectively carry out what the Russians term 'Konversiya', turning military capacity into consumer goods capacity – and do so competitively – remains to be seen.

In the next chapter, we move on to discuss the impact of changing markets and technologies on skill-formation, and how managements coped with the challenge.

Chapter 2

Markets and technologies

This chapter summarises the arguments which led up to the start of the project. First, it describes the changes in technology which have been associated with advanced manufacturing systems and which have led to greater flexibility, product customisation and so on. It then looks at the human resources management implications of such changes in the workplace, and amongst others covers the impact on the occupational structure, the level and the mix of skills. Finally, it sees how these changes affect the diversity and cost of training and their implications for selection and level of recruitment. Over the last decade or so, technological changes have altered the nature of the manufacturing process more than at any period since the end of the last century at least. Some continue to speak of a 'third' industrial revolution. Comparisons have been made with earlier changes described by Adam Smith in *The Wealth of Nations*, first published in 1776.

The impact on the world of work cannot be exaggerated, whether acting through technology, organisation or markets. The way in which the daily lives of ordinary citizens are affected may be manifold, as the new technology affects not only production but consumption, welfare, leisure and many other aspects of social life. The scenarios for future socio-technical change which have been put forward range from the optimistic to the pessimistic, but microelectronics is not the only factor affecting the economy. It is hard here to make 'scientific' statements about the long term, short of extrapolations of the present. Speculation on the industrial sociology of the future however is hard to resist. It seems likely that the world of stable, specialised mass markets is being eroded by new changes, the social consequences of which are only now becoming clear to the informed observer, let alone the lay person. The impact of microelectronics on the work force will probably occur through changes in the labour process. What will, for example, be its effects on human resources strategy?

CHANGES IN MARKETS AND TECHNOLOGY

In order to remain competitive, many firms are investing in new technology incorporating microelectronics because markets are becoming more complex, and possibly more differentiated, after the environment in which firms operate has become more variable and uncertain. Thus, market shifts may be seen as a plausible trigger for many changes in how goods are produced, and the shape of the production function. Even if the technological state-of-the-art is often ahead of the market, new manufacturing technologies have opened many marketing directors' eyes to new markets.

International competition, saturation of existing demand, or more sophisticated consumer tastes as incomes rise in advanced economies present new challenges to producers. It is clear, however, that whether or not Adam Smith's dictum that 'the division of labour is constrained by the extent of the market' still applies, that the so-called 'microelectronics revolution' has affected the market–technology–product nexus. But one may question the degree to which 'the extent of the market' leading to conventional economies of scale is still the crucial constraint on the division of labour within firms. As the new technology has become more advanced, it has also become substantially more flexible than previously.

> Manufacturing flexibility means a variety of different things to different people in different situations. It is often viewed as the ability to easily adapt the manufacturing processes to (a) changes in the product design, (b) changes in volume or batch-size – including the ideal of a batch of one, (c) alternate materials, (d) minimize set-up times, including the ideal of zero set-up time. Flexibility in manufacturing refers as well to equipment (dedicated vs. numerically controlled or computer controlled), to people (narrowly trained vs. multi-skilled) and to systems (e.g. developing a computer-controlled parts categorization system to minimize changes to production processes when product variations are required. This is often a prerequisite stage to developing a Computer Aided Design/Computer Assisted Manufacturing or CAD/CAM system).
>
> (Kolodny 1985: 208)

By combining the benefits of flexibility and specialisation, the new manufacturing systems represent a quantum shift in production techniques, with important implications for labour – as well as product – markets, as we shall soon see. Such systems require flexibility in organisation and labour, as well as in technology. Of course, flexibility was never absolutely absent in former times but the difference may lie in the degree of relative flexibility available with the new developments. Such flexible production systems are now able to customise products rather than offer standardized runs as was the case with the conventional economies arising from secondary mechanisation. The new production methods rely on the 'economies of scope', rather than of 'scale'. Without the microelectronics-based technology available in the last decade, such methods would not have been available.

Table 2.1 Classification of technological stages

Mode of organisation	*Flow production*	*Technology*	*Division of labour*	*Product*	*Market*
1 Small workshop	Absent	Flexible	Low	Customised	Differentiated
2 Early factory	Nascent	Relatively flexible	Medium	Relatively standardised	Relatively differentiated
3 Modern factory	Present	Relatively inflexible	High	Standardised	Homogeneous
4 Automated plant	Modified	Flexible	Modified	Relatively customised	Differentiated
5 Manufacturing cells	Responsive	Flexible	Negotiated	Customised	Rapidly changing

Source: Warner 1986: 281

A diagrammatic representation of how stages of technological and organisational change have evolved to the present state of the art is set out in Table 2.1. The scheme has affinities with other recent attempts to develop a theory of work organisation which takes into account the advent of automated production-systems, such as that centred on the notion of 'flexible specialisation'. Based on a study of North American, West German and Italian examples, the authors Piore and Sabel (1985) point to examples where small and medium-sized companies have distinct advantages over large enterprises. The long-term consequences for employment and its location are considerable; but in terms of microeconomic considerations, the implications point to less rigid organisational structures, up-grading of skills and so on. In a nutshell, new technology can render 'small' as 'beautiful'.

Technological change also seems to be leading to the possibility of locating firms in areas other than the old industrial heartlands. There is a direct link between the closing of large units in the traditional industrial regions, and the emergence of small units elsewhere (Shutt and Whittington 1987).

This has affected the economic balance between large and small towns:

> the re-emergence of market towns means that a definite change in the social and economic lifestyle of Britain is taking place. More and more businesses, especially in new high-technology industries like computing, are being attracted to such manageable country towns – and the process further depletes the traditional, and unwieldy, urban industrial centres like Glasgow, Sheffield and Liverpool.
>
> (Beresford 1986: 22)

Discussion of the role of technology frequently becomes immersed in conflicting definitions of determinism. It may well be the case that technology does present the 'small is beautiful' approach, with smaller more skill-intensive units competing with the less flexible mass-production giants, as more viable than before. However, as the protagonists of flexible specialisation themselves assert, it is social actions rather than technologies that determine the shape of organisations – the view that 'mechanization has as its precondition and as its consequence mass production' is, after all (they state), a 'restatement of what happened, not the summary expression of an inevitable logic of interest and efficiency' (Sabel and Zeitlin 1985:133ff).

In effect, a shift has been occurring away from hierarchical organisation (in the sense meant by Williamson 1975) towards 'market' principles, with firms gradually breaking up into trading units. This process offers flexibility but may tend to fragmentation in strategic approach.

CHANGES IN OCCUPATIONS

By using the new technology, firms are finding that older production rationales are radically changing. The 'previous logic of socio-technical design has been geared to specialized homogeneous mass markets', based on 'inflexible automation, an erosion of craft worker skills, and increased emphasis on separate planning activities' (see Sorge, *et al.* 1983: 158). This is now changing as small-batch production becomes viable at something nearer to mass production cost-levels and new patterns of skill utilisation.

As products are customised using flexible technology, the ways in which the machines are used affects the occupational profile in the workplace. In order to best utilise the technology available skilled operatives often become more predominant, backed up by highly trained technicians and graduate engineers. An 'holistic' approach with reduced division of labour is being found to be more efficient than traditional 'Tayloristic rationalisation' patterns (Schumann, 1990).

The British engineering industry provides a good example of how technological and economic changes have altered the balance of the labour force, to give proportionately greater weight to upper strata of the occupational structure. For example, in the Scottish electronics industry no more than 50 per cent of the labour force are manual workers (Henderson 1989: 127). While facile prognostications of robot-operated factories may be unjustified (Senker and Beesley 1986), there is sound evidence of deep-seated changes in occupational structures, at both enterprise and societal levels. There is little doubt that trends in the electronics industry point the way to changes which will set the pattern for industry as a whole. Not only will fewer people be employed in manufacturing, but those who are will be different in kind than hitherto. Within the electronics and computer-related sectors, we are already moving towards a 'white-coat' labour-force profile. As more electronics are introduced into products, original manufacturers (such as those of machine-tools) are less capable of supplying products, so equipment suppliers take over the marketing. The workforce of the original manufacturers become relatively less skilled, whilst that of the suppliers becomes more skilled.

Several plants we visited during the present study had limited numbers of employees other than those at technician and graduate levels of training. The annual intake of graduates to high-technology firms in these cases now constitutes an increasing proportion of their recruitment. A 'mushroom-shaped' profile is increasingly to be found in their occupational structures. A small number of managers in these firms direct a much larger number of graduate engineers supported by a much narrower group of technicians and below them a diminishing number of 'other ranks'.

Commenting on the labour requirements of the computer manufacturing industry in the 1980s, one observer noted that:

> Semi-skilled and unskilled labour has been eliminated; the ratio of technicians to craftsmen has increased from less than five to one, to nine or ten to one; and the ratio of scientists and technologists to technicians has increased from less than one in five, to better than one in two.
>
> (Bell 1984: 74)

Should these assumptions be projected, ninety-five jobs in a hundred could go at shop-floor operative level, with the remaining five concentrating on 'supervision, fault diagnosis and rectification ("trouble-shooting") and maintenance, requiring for the most part training of a technician type'.

Microelectronics has been the breakthrough responsible for one particular form of job displacement, as it has greatly cut out the labour-intensive wiring-up and assembly operations. Miniaturisation and automation have gone hand in hand with large-scale integration (LSI). Robots and microcomputers link stages of the production process, with high degrees of complexity, adaptability and flexibility of operating cycles.

CHANGES IN SKILL LEVELS

Changes in skills mix are likely to occur in the workplace in two ways: first, there will be employees with distinctly new skills represented in the workplace, and second, because many workers will have several new skills mixed with their existing ones. It is clear that, given rapid technological change, there will be 'hybridisation' of skills (Campbell and Warner 1987). Boundaries between craftsmen and technician's work could blur with a need for more broadly trained engineers who are able to adopt a systems approach to implementing the new technology. Rather than the trend to specialisation of function continuing to a higher degree as was the case with industrialisation and technological change in the past, the application of microelectronics seems to imply the reverse. However, technology *per se* 'does not determine the skill mix required'. (Williams 1984: 212).

Technological change generally may, however, make the manufacturing workforce smaller and more skilled, but skills may shift in emphasis and focus. For example, greater expertise may be needed for printed circuit board assembly. With the introduction of microprocessors, small-batch production means a greater emphasis on trouble-shooting skills at the craftsman/technician level. It is also likely that 'the single disciplined craftsman has no foreseeable future in most front-line maintenance situations'. The shift to a multi-role/multi-disciplinary craftsman, in this view, reflects 'the general shift from a differentiated and specialist based organization to an integrated one' (Cross 1985: 193).

With a greater skills mix in the workplace, more workers are able to cope with a greater variety of tasks both in operations and in maintenance. For example, the more production systems are automated, the greater the need to keep such equipment in continuous operation and prevent breakdowns. Experts advise that it is 'unwise to rely on automatic diagnostics to deskill maintenance and to remove the need for training. Successful implementation of automatic diagnostics demands new training programmes' (Senker 1984: 244). In the examples cited, this account describes how electricians have been instructed on how to program Programmable Logic Controllers, and fitters shown how to use Electronic Diagnostic Equipment for locating faults, for which they need training in pneumatics and hydraulics.

Multiple skills are increasingly in demand; for example, technicians often work in common design and development groups, and have to co-operate across occupational boundaries, with theoretical knowledge in a range of fields matched by practical and diagnostic skills. Another important development is the polarisation of skills between technicians doing non-routine and routine computer-assisted work. Mixtures of operator and maintenance skills are also needed at sub-technician level, to part-program CNC machine-tools for instance, as well as for coping with faults or breakdowns. Mechanical and electrical/electronic skills in future will need to be combined. Some would go so far as to say that it is probable

> that workers will have a responsibility for a longer span of production equipment because there will be greater system integration and fewer workers manning the production equipment. Many workers will merely carry out a monitoring role – patrolling and inspecting a system when all appears to be functioning properly. However, he or she will be expected to react sharply to an incipient crisis and take

> corrective action when anything goes wrong. This may involve knowledge and competence of a highly integrated system of machines.
>
> (Gill 1985:74)

An example in our own fieldwork in a large firm making information-technology equipment in the electronics industry, suggested that in this plant, about half of the staff (and half the technicians) were trained in broadly based electronic expertise. In R&D, most graduates had been trained in electronics hardware or software. Most of these entered the development area, where a majority of those involved were graduates. Employees may learn a wider range of less narrow skills ('horizontal' skill acquisition) and/or more or fewer more specialised ones ('vertical' skill acquisition) – (see Figure 3.1, p. 29).

Maintenance craftsmen may need to have a wider mix of 'engineering' and 'electrical skills' (horizontal) as well as new electronic technician skills (vertical) for example (Rothwell 1984: 128). As the hybridisation process proceeds, more diverse training is needed to cope with the increasing range of knowledge to be imparted to the labour-force. It is possible that 'societal' variations, based on national differences in educational and training institutions, and established work practices may be visible here. As the technology becomes more complex, the training requirements may become broader in scope, but this may emerge as broader in say (West) Germany compared with Britain.

Many firms do not yet appreciate the need for such systematic training. As Rothwell has pointed out:

> The need for extensive training in new technical engineering skills for maintenance and process technicians and craftsmen was only just beginning to be appreciated. Suppliers' own courses were rarely adequate to cover long-term needs and reliance on suppliers' own maintenance technicians (who might have to be flown from Germany) was even less satisfactory.
>
> (Rothwell op. cit).

At present, most companies in Britain rely on traditional apprenticeships, HNCs and electronic and software degrees. The microelectronics component in these courses is par for the electronics industry but will grow in future years. The shop-floor operatives receive mostly on-the-job training, with conversion courses for field-service engineers. Training at shop-floor level needs greater emphasis. Many companies have insufficient budgets for in-house training. Sometimes they send supervisors on customised courses at a local polytechnic and apprentices to technical colleges on day-release, in addition to the minimal in-house training they organise. One company with 650 employees in the office automation-assembly field had an annual budget of £300,000 for in-house work, however.

In order to impart the range of skills needed, such diverse courses require longer periods of training, even if particular parts of the course can be more effectively learnt by user-friendly computer-assisted learning methods. Rothwell and Davidson have noted

> 'Conversion Training' was the title most commonly given to what might most easily be understood as 'new technology' training. In 'information system' applications, it usually included keyboard skills but this was seen as less important than the use and understanding of computer files and how they relate to each other, because of the importance of maintaining a highly accurate common data base. It usually consisted of some advance demonstration and practice followed by on-the-job coaching once the system went live. Sometimes it was followed up with some 'mixed' group training to try to assist employees' understanding of the relationship of their work (or mistakes) to other sections.
>
> (Rothwell and Davidson 1983: 118)

Training for robotics, for example, requires a range of background skills, including, in some British training courses, ten days of elementary electronics, twelve days of microprocessors, and four days respectively on pneumatics, hydraulics and fluid logic. In addition, logic-theory involved in electronics must be included, hence practical use of maths systems, with binary hexadecimal and Boolean algebra (Tarbuck 1985: 79).

It is likely that a split may occur between large firms on the one hand, and small firms on the other. Large firms will be able to offer training on an adequate scale in-house, but small, and even medium-sized firms may not. As far as shop-floor personnel were concerned, there was no training policy at all in most companies, with no involvement whatsoever by trade unions. With longer courses, the cost of these goes up and the opportunity-cost rises if workers are absent from the production process itself. The average length of training for operatives ranged from a few days to a few weeks. Customised training was often done by the manufacturers, consisting of two-week modules in many of the British cases investigated. Sometimes the training was done by the company itself to save time and money. Manufacturers often used the need for no training as a 'selling-point', or included a training package as part of the sale, with for example, CNC and CAD systems.

Another reason why training may be inadequately provided for, in terms of company-resources, is that the workforce is often seen as static. Or, alternatively, the needs of a given level of employment in the firm may be viewed statically, without foresight regarding the need for enhanced levels of expertise, in-house skill building and so on. Again, insufficient attention may be given to the relative costs of in-house versus external training, and how these two categories fit into a human resources development strategy (see Campbell *et al.* 1990).

The absolute length of training is less a problem than the need for increasingly frequent up-dating of ideas as existing knowledge becomes more rapidly obsolete. According to the then Manpower Commission chairman, Mr Bryan Nicholson, 'Between half and three quarters of Britain's workforce will need to update their skills and knowledge during the next five years just to keep pace with new technology.' In the early 1980s British firms spent only 0.15 per cent of their sales turnover on adult training, whereas a high proportion of American counterparts disbursed 2 per cent or even 3 per cent according to a Manpower Services Commission sponsored survey of 500 firms in the United Kingdom (cited in Warner 1985:).

CHANGES IN SELECTION

As training requires greater investment by employing organisations, as we shall argue in detail shortly, they will be more selective in recruitment to find employees with the most appropriate prior qualifications and aptitudes, and to minimise the risks of selecting those who will not be good long-term 'training investment'. Many firms have long been reluctant to carry the costs of training people, therefore the shortages of skilled, technical, and maintenance manpower. These include lack of suitable candidates for apprenticeships, dismissal of skilled staff early in the recession due to rationalisation, inadequate training schemes, demarcation practices, etc.

Recruiting is one thing, holding on to graduate or technical trainees is another. For example, in one British plant studied, graduate intake for the plant (all engineers) was fourteen in 1985, compared with twenty-five for 1984, out of the total of 330 employees. The turnover per year, however, was over 20 per cent per annum. With the skills shortage, labour turnover thus becomes a problem. Software engineers are at a premium and only stay a couple of years in many cases. If promoted to senior software analyst, for example, they might stay. But often salaries are not raised to market levels for fear of upsetting differentials in the company.

When companies were expanding, there was less chance of a skill and rewards problem. With expansion, there was less need for redundancy. Indeed, there was a shortage of skilled labour particularly at the top end of the skills range, where some firms would be happy to pay the 'market' price to recruit the qualified personnel they need, especially software engineers and systems designers.

Selection criteria will ultimately affect organisation structure in turn (at least indirectly), although there is insufficient space here to discuss this in detail. It is clear that the types of people recruited will be determined by the existing organisational structure, but both they and their training qualifications will eventually affect its shape. If firms begin to look increasingly closely at new trainees for longer-term employment in the 'core' labour-force, they will be more likely to only take on the minimum they need, particularly as the costs of training grow for the reasons addressed earlier.

If all we have described above comes to pass, training costs will rise relative to other employment costs. With new technology, this disincentive to recruit will also be additional to the capital cost per job, with the existing relatively high capital–labour ratio.

With Flexible Manufacturing Systems (FMS), each worker may be responsible for several millions of dollars or pounds of equipment for example. While it may be argued training costs should be a low percentage of total costs, it is reported that the former could be between 5 per cent and 10 per cent of the total (Thompson and Scalpone 1985: 226). The human resources investment supplements the physical capital cost. There is however the following paradox of human resource investment, for the full effectiveness of physical capital investment may not be realised if skilled labour is not deployed and trained appropriately, as was apparent in our earlier Anglo-German study of CNC applications (Sorge *et al.* 1983).

There was a trend, however, to changing the workforce by recruitment rather than training. But the smaller the firm, the more likely they are to train their people themselves and upgrade their skills, because they are unable to afford to bring in more expensive graduate trainees. Larger firms are increasingly looking for manpower with 'paper qualifications', more or less. Some firms have tried to dispense with operators altogether: 'unmanned' and reliable processes are designed and put into practice. But as Peter Senker has pointed out:

> In many areas, engineers' efforts to dispense with workforce skills are only partially successful – and this partial success may often result in a worse situation than if the need for workers had been accepted and the equipment designed around the availability of a trained and trainable workforce.
>
> (Senker 1985: 161).

An alternative strategy is to create a homogeneous cadre of skilled employees with mixed skills – all tasks can be carried out, whether workers are tackling electronics, programming or other related operations in advanced manufacturing.

ONGOING TRENDS

Training is increasingly being recognised as a strategic area and a key source of competitive advantage. Programmes are being aimed at generating a workforce capable of continuous learning and participative change (Benton *et al.* 1991, Wickens 1991). Broadly educated managers as well as workers have distinct advantages over narrowly trained ones in this respect. These may, however, have broadly defined capabilities in terms of in-house skill requirements, as opposed to labour market-oriented training, which in turn could enhance their potential for seeking employment elsewhere.

It can be seen that one implication of the argument put forward in this chapter is that recruitment will be more selective as a result of the process set in motion by the change in the market–technology–product nexus, as we shall see in the model to be presented shortly (see pp. 28–33). It is not only that job losses will occur because of the initial desire to achieve productivity gains, but that even if these do not necessarily occur when new technology is installed, the possibilities for enhanced recruitment of the skilled employees may be constrained *ceteris paribus* because of training costs involved, as the model implies. Even if product markets are expanding, and firms can compete for example, by customising their products, the labour-market implications may not be as positive as they might have seemed to have been.

On the other hand, it is clear from the arguments and evidence presented above that while there will be fewer employees in the workplace, we would hypothesise that the average level of skills of those remaining in employment may rise. Moreover, that there will be a lesser division of labour due to the eventual emergence of the 'hybrid-skills' phenomenon and therefore by implication an increase in the level of skill integration across-the-board in the enterprise. It is important to stress here that attention should not be exclusively focused on the semi-skilled and skilled operative (as is done in so much of the literature on the labour-process), but all the way up the occupational structure or hierarchy in the firm. The categories below, but not including craftsmen (and leaving out canteen-workers), constituted just over 45 per cent of the employment in the British engineering industry over the last decade. The levels above this group amounted to just over 55 per cent of all jobs, with the greatest rate of increase in the latter group as a whole.

A well-known North American authority on the employment implications of robot-technology has called the above phenomenon the 'skill-twist', as follows:

> Perhaps the most remarkable thing about job displacement and job creation impacts of industrial robots is that the jobs created and the jobs displaced do not match up very well. The jobs eliminated tend to be semi-skilled and unskilled, while over 50 per cent of the new jobs created require a significant technical background. That is what we call the *skill-twist*, and we think that the skill-twist is the finding of our research. In fact, we submit that the skill-twist is the true meaning of the so-called robot revolution.
>
> (Hunt 1984: 14)

Whether a two-tier structure is inevitable is debatable, in many applications at least. The old multi-tier structure may go, but what replaces it may still conceal implicit hierarchies, depending on the societal complexity involved. Greater degrees of freedom may be more likely than argued by the technological determinists. A three-tier structure may be a better prediction, with a middle-zone between the two extremes and with greater flexibility between tasks.

Again, the argument concerning the rise in the level of skills of those remaining in employment in the workplace may sometimes be considered subjectively, especially as far as any individual or sub-set of employees (or their trades-union) are concerned. If this line of argument is pursued, it may be hard to say if the average level of skills is higher *per se*, although the average level of training (defined in terms of existing training qualifications) represented in the firm may rise amongst those in the workforce at that particular time. The gap between the demand for highly skilled labour, and the supply of indiscriminately qualified labour partly results from the rapid pace of technological change in key sectors, not always fully anticipated by employers, trade unions and goverment agencies, and partly from the effects of recession. There is little evidence that the gap is being bridged, or that the situation, at least in Britain, is improving in spite of the lengthening dole queues (see Keegan, 1990).

Given the exclusion, if present trends continue, of so many from the manufacturing labour-force (or from employment generally), because of lack of skills as outlined in the above examples, it becomes imperative to expand the training programmes available to those excluded on a massive scale. This may be justified on economic as well as social grounds, as if anything the relative short-fall in competence is increasing faster than the growth in technological sophistication.

While there are both skills shortages specifically, and jobs shortages generally, these trends need not persist indefinitely if there is much greater public investment in training. In any event, human resources investment is still a fraction of physical capital investment if it is considered at the macroeconomic level. A British training manager recently produced an apt phrase, 'Training is cheaper than ignorance.' True as this is at the enterprise level, it is no less valid when one looks at the costs of inadequate training macroeconomically. In many cases, management perceive the problem as symptomatic of the electronics industry, indeed it may well be true of the economy as a whole.

When the recession of the late 1970s occurred, it was assumed in Britain that all problems concerning skills would involve a question of a surplus not a shortage. In some respects this was true; companies vastly accelerated their programmes of restructuring, and in many cases, automation, all of which led to a massive reduction in the industrial workforce. Progress, 'restructuring' and technological advance also provided a smokescreen for the collapse of much of Britain's manufacturing industry. Decreases in the numbers of technicians and draughtsmen employed nationally have more to do with the collapse of whole sectors of industry than with technological displacement (MSC 1985). A smokescreen was also provided by the way in which, as we have described, these events fitted neatly with the view that manufacturing was 'old' and would be replaced by 'new' services. The element of truth contained in this prevented people from noticing that, of the advanced countries, only Britain saw significant absolute (as opposed to relative) decline in manufacturing in favour of services. That such an absurd view could have held any ground, we believe, is proof in itself of the crucial importance of national belief-systems, and their ability to override logical thinking.

This period did however see a growth of awareness, by the way of reaction, of the extent to which the key role of engineering in the country's prosperity had been undervalued and ignored (Finniston 1980). This was succeeded later by an explosion of publicity over the 'skill shortage' and a more generalised concern over the resources being allocated to education and training in Britain compared to overseas competitors. Although Meager (1986) maintains that there is no proof of a causal link between skill shortages and economics performance (one might indeed be a symptom rather than a cause of the other), many studies have pointed to serious deficiencies in these areas, particularly where 'information technology' skills are concerned. According to recent estimates, there are at present over 200,000 professionals in the IT field in the United Kingdom. Only 8,000 a year are produced at the graduate level each year, however, which is nowhere near enough to meet the country's needs over the next few years (see Pearson *et al.* 1990 for further discussion).

As if to emphasise the extent to which new technology is a Janus-faced phenomenon, taking more from one group and needing less from another, another influential study (Northcott 1986), estimated that 80,000 jobs had been eliminated directly by the application of microchips over two years in Britain. In the long run, the Policy Studies Institute advised, some macroeconomic action needed to be taken to deal with the gap between jobs created and jobs destroyed by new technologies (Northcott 1986). Government intervention (if this is what 'macroeconomic action' implies) on this account has not been especially forthcoming. In the meantime, approaches had been made to the Department of Trade and Industry by large companies complaining of shortages in the number of electronics hardware and software graduates. These complaints led to the formation of the Butcher Committee, whose report (DTI 1985) used company sources to predict the supply and demand of such graduates and to examine how the responsibilities of government, industry and the education system should be defined. The Butcher Report, with some justification, put the bulk of the responsibility onto employers, who were expected to improve their links with the education system in a variety of ways, and not to sit back and wait for government or other companies to reverse the position (Press Release, 12 September 1984).

The implication was that much of the problem lay in British employers' attitudes to training, even if the supply of university-trained recruits was inadequate. This point had already been made by Albu

(1980), who described how only a minority of British firms accepted the training of engineers, contrasting the position with that of German industry. Similar conclusions were also reached by Wagner (1983) who emphasised the ways in which the German economy was assisted by the comprehensive training it offered at intermediate level, again in contrast to Britain. Far from 'over qualifying' workers, the breadth of the training given in Germany made workers more adaptable to new jobs or to technological changes in their existing jobs (Rose 1990a and b).

The most comprehensive criticism of British employers' attitudes came with *A Challenge to Complacency* (Coopers and Lybrand 1985). Notable among the recommendations covered was that which argued that, since companies devoted proportionately far greater resources to management development than to training generally, then management development courses should be geared to 'training about training', since lack of management awareness about the need for training at all was a major part of the problem (see Campbell and Warner 1990).

Two shifts in awareness, we would argue, have begun to occur as a result of these, and many related studies. First, shamed by the much-quoted statement that British companies spend so little of their turnover on training (Gordon 1985), companies are beginning to undertake more ambitious training schemes, often in collaboration with educational or governmental institutions (Fudge 1986 provides several examples). As a result, training departments are gaining in resources and status, particularly since recruiting itself may be significantly helped by the training opportunities offered to prospective employees, including graduates. Second, the shortage of graduates has had another effect in that companies, whilst still complaining about graduate shortages (which may in part be due to 'setting the sights too high' – Jenkins and Vandevelde 1985), are seeking to overcome them through in-house training (see Tarsh 1985). These developments may to some extent lead to a partial reversal of the elitist view of technological change and skills that has dominated British thinking, leading companies to assume that all technical innovations should reduce skill requirements below graduate level (Senker and Beesley 1986).

Finally, where both graduates and technicians are concerned, the trends in product-technology that have led to an increased requirement in hybrid or product-specific skills (the two are related in many cases) appear to necessitate a far higher degree of in-house training, at least in the more sophisticated sectors of engineering (see Campbell and Warner 1989b, 1990)

In the next chapter, we shall move on to outline a model of new technology and skills and attempt to conceptualise the causal connections involved.

Chapter 3

A model of new technology and skills

In order to best utilise the technology available in the current context, manufacturing firms do appear to be placing a greater emphasis on higher skill levels, on the shop-floor and elsewhere. This change does of course coincide with serious reductions in the total numbers employed, with the less-skilled losing out. At the same time, there is a growing trend to return the focus of skills back to the shop-floor. Innovations such as CAD/CAM remove much of the need for expanded white-collar control groups away from the shop-floor, although a management may retain them (unnecessarily) as their philosophy so dictates. The trend, whereby communications and involvement between engineers, programmers and technicians and the operators are increased, contrasts with the trend under mass production conditions and technology for skills to be removed as far as possible from the point of production. The return is more marked in Germany than in Britain. Naturally, there are already counter-trends to this movement back to the shop-floor. The rule seems to be that if an organisation's culture and management strategy demands a perpetuation of mass production styles of working, and sees even the most flexible technology as a means to this end, then control mechanisms will be made more constricting, more centralised to the extent that the technology is potentially flexible.

Is new technology, then, the crucial determinant of workplace organisation after all, or do management strategies determine all? Is 'flexibility' – whether in skills, training or deployment – a significant, as well as a much-abused, concept? Who gains as a result of these trends? Are market and technological changes 'considerably enhancing managerial control' (Gill 1985: 32).

In Figure 3.1, we attempt to conceptualise the causal process. Trends are grouped in pairs where they coincide, and these are set out schematically.

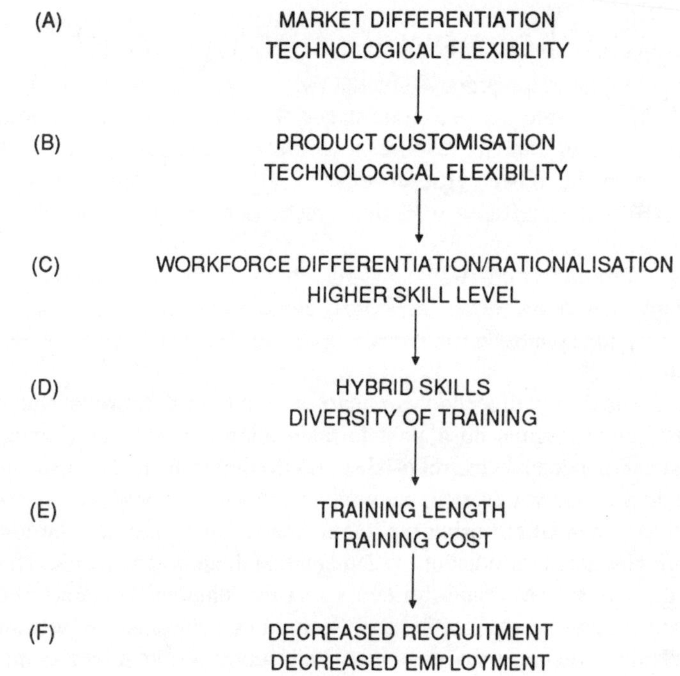

Figure 3.1 Changes in markets, technologies and skills in periods of slow economic growth

The model begins with trends arising from slow growth in product markets (A), the concrete application of these trends (B), then, initial effects on the workforce (C), the bringing of skills and training policies into line with the perceived needs of the market/technology nexus (D), the resulting growth of training activities and costs (E), and finally the effects on the level of manufacturing employment (F), with recruitment becoming a prelude to, rather than a replacement of, training and re-training, and thus involving a more conscious investment on the part of employers. The result may, in the absence of a significant increase in aggregate demand, be a further decline in overall numbers employed, with manufacturing employment becoming increasingly an elite occupation. This change may coincide with an even sharper division between core and peripheral workers, many in poorly paid service occupations. These last, less desirable conclusions, whilst already perceptible and indeed already advocated in some quarters, lie outside the scope of our study. The earlier points will be discussed with reference to our findings.

According to the proposed model (which develops an earlier version – see Warner 1986), six separate stages of a process linking technology with a set of human resources variables may be extracted. The phenomena have been grouped together where – particularly with (A) and (B) – it is difficult to assign precedence or a clear direction or causation.

(A) From the mid-1970s onwards, the availability of flexible technology, based on microelectronics, coincides with a crisis of mass production and a trend to diversity in the market, as described at length earlier.

(B) Flexibility of technology enables the meeting of a wider range of specific customer needs/demands than before. At the same time, as customers become aware of the potential of the technology, they demand a wider range of functions from a given product. As more functions can be encapsulated into a smaller space through the use of microelectronics, so customers demand that the newly available 'space' in the product be filled with new or more elaborate functions (this is more clearly the case with such electronic products as the telephone exchange, which in the past was much larger and involved complex electro-mechanical controls). With more and more functions enclosed in a small space, the technology, both in hardware and software, becomes increasingly complex.

(C) As this process continues, the workforce is 'rationalized'. The removal of many of the production steps associated with electro-mechanical assembly leads to a decline in the number of semi-skilled jobs. Even where no reduction occurs, the proportion of employees engaged in such work will fall, the numbers failing to increase in line with increases in turnover or profits. At the same time, more professional specialists will be recruited to cover the increasing load placed on design and development. Matrix forms of management and on-the-job autonomy may be introduced. These activities will begin to outweigh production in time, money and perceived importance. In many cases, notably those where product-technology is highly sophisticated, production may be 'hived off' either through increased buying in of components or through sub-contracting the manufacture of sub-assemblies. A more flexible form of task organisation characterises this stage of development.

(D) As products become more complex the interfaces between different design or production disciplines become more tangled and integrated. A similar development may be encouraged regarding the skills of employees. Professional specialists may need to take more notice of the input of other disciplines or to take more of a 'systems view', with greater awareness of the product as a whole, rather than just their particular contribution. Equally those involved in the assembly, testing and maintenance functions may require a more general understanding than before, and possibly the command of a wider (if not necessarily 'deeper' range of skills) which may in turn require regular up-dating. The same is true for specialists, who particularly where computing skills are concerned, may find their knowledge to be obsolete surprisingly early in their career. As a consequence of these trends, employers may need to invest in a wider range of training activities than before. Even if they fail to recognise this need, they may find it increasingly difficult to attract new specialist recruits without providing the 'career insurance' of retraining at regular intervals.

(E) All this should lead to the expansion of the training function, with a larger amount of investment for a larger number of days per year to be assigned to training each employee. In higher technology companies, as product-technology increases in sophistication more 'state of the art' training, involving close co-operation with academic institutions at all levels, may be needed, all of which will entail greater cost per head as well as greater administrative effort. Rapid technical advances across several sectors of industry usually lead to skill shortages and Britain has been particularly prone in this respect. With recruitment of people at many levels of skill becoming a problem, companies may find they have to devote even more resources to developing the skill base they already possess.

(F) Extrapolating from all the above trends (frequently encountered in our fieldwork), it follows that recruitment of an individual becomes more of an 'investment decision' than before. Fewer people may be chosen, and more carefully. The result may well be a much smaller, more highly skilled workforce in manufacturing, which may have a less than beneficial, or even a negative effect on rates of employment. Although increased competitiveness by key companies may have spin-offs for employment elsewhere and increase aggregate demand, the evidence for this in Britain appears to be less than encouraging. Viewed as a whole, then, the process (conceptualised in Figure 3.1) suggests that up-skilling and a greater emphasis on training, whilst benefiting a core of workers in industry, may entail increasingly concrete 'caste divisions' between the above and their 'peripheral' counterparts.

Questions of technological determinism inform any debate in this area. While some of the classic contributions incline towards a heavy emphasis on technology, others assert that technology has little impact beyond the purposes of those reponsible for applying it, although this tends dangerously towards a concept of managerial omnipotence or omniscience. Others take a middle view, seeing the advantages of innovations as not automatic and not necessarily rapidly achieved. Sociological discussion on related themes has stretched from micro-level studies of job design, to macro-level investigations of the implications for social stratification.

The debate on automation has continued since the 1950s with varying degrees of intensity in recent years. Whether or not it is possible to develop an integrated theory of work organisation which adequately explains current developments, or even what such a theory would look like, is still hard to predict. At present, the situation in the theoretical domain is less than satisfactory.

In order to explore how skills are affected by developments in new technology, a set of research studies were initiated in Britain (with parallel studies taking place in West Germany). Semi-structured interviews were held with engineering, personnel and training managers at a variety of manufacturing locations where micro-electronics had been incorporated into products. Documentary material provided by the companies was also used, where possible. The aim was to examine possible links between skills and training strategies and the technological developments taking place in product lines (including changes in technical complexity, batch size and customisation). Such connections, it is hoped, may point to a satisfactory explanation of the directions being taken in the areas of work organisation.

BACKGROUND TO THE STUDY

The present study is in many respects a sequel to an earlier research project, *Microelectronics and Manpower in Manufacturing* (Sorge *et al.* 1983). The authors found a number of important national differences concerning the deployment of CNC technology in Britain and West Germany, differences which varied according to the size and batch-size of the plants being investigated. German firms placed a much greater emphasis on craft and 'intermediate' skills (a higher incidence of operator programming for example) than did British firms. Whereas at the small firm level differences in this regard were minor, at the large firm level they were much more significant, particularly where the production of small batches were concerned. British firms interpreted their skill needs according to a mass-production perspective, geared to economies of scale and mass homogeneous markets, this outlook appearing to influence decisions even in the small batch setting. Furthermore, this mass-production perspective was interpreted in Britain along the lines of what the authors termed a 'services logic' (as opposed to the 'industrial logic' which characterised the German firms). That is to say that technological developments in Britain were perceived as moving towards complete automation, presided over by staff specialists. Progress thus meant a 'flight of skills' away from the factory floor and direct production. In Germany, on the other hand, staff specialists tended to move onto the shop-floor where they interacted with craft labour, the latter being seen as constituting the nucleus of operations.

The authors concluded with an attack on the way technological change is perceived, particularly in Britain, as leading to a 'post-industrial' or 'information' society, a perception which, they contended, had less to do with actual technical developments, and more to do with cultural biases and traditions. In this way, they argued, 'pure' information-related skills are seen as more important than the substantive skills concerned with the industrial processes to which the 'information aspect is ultimately peripheral'. They thus demanded a repudiation of the 'services logic' and a renaissance of the 'industrial logic' (see also Campbell *et al.* 1989 for further discussion).

TECHNOLOGY, DEMAND AND EMPLOYMENT

Will the changing technological and market relationships described above create or destroy jobs? Opinions divide into optimistic and (in contrast to the eighteenth-century classical economists) pessimistic scenarios. Adam Smith in his time would have been most surprised at the current concern regarding shrinking employment. As an appreciation points out

> it is essential for the understanding of the dynamic mechanism to realize that in this conception technology, and in particular the introduction of machinery, is regarded as a complement of, rather than a substitute for, labour. In other words, far from displacing labour and thus exerting a potential pressure on employment and wages – the major variable in Marxian dynamics – the division of labour in this inclusive sense is itself conditional on a prior increase in labour supply.
>
> (Lowe 1975: 417).

It is, however, very hard to make 'scientific' statements about the long term, short of straight extrapolations. Microelectronics is only one aspect of the technological changes affecting the economy, but perhaps the most topical and newsworthy, and often introduced for a wide range of purposes.

Child has observed, using an organisation theory perspective, that: 'Managers will normally have several goals in mind when introducing new technology into companies' operations. The emphasis between these is likely to vary according to the priorities and purposes of their organisation and the context in which it operates' (Child 1984: 213).

To cite some examples: robots may specifically be substituted for direct labour in the car industry; word processors may partially replace some labour inputs; desk-top minis may not replace any labour, but enhance the manager's ability to do his accounts more effectively. In a large company, some labour may be shed (as in the case of a mail-order company which recently cut its workforce drastically). In a small company, the only effect may be that new employees may not be taken on. If new up-graded jobs are introduced through VDUs, there may be less need for supervision so roles may be reallocated but the employee may remain on the payroll. Advanced technology may simplify/or elaborate management organisation (see Campbell and Warner 1990).

Other writers take a more sweeping view. For example, from the economists' standpoint, Peitchinis (1983) traces the development of microelectronic technology over the past twenty-five years as the first stage of a three-phase evolution where, to start with, the effects of stand-alone computers saw an expansion in the number of jobs available. In the second phase, these stand-alone instruments and processes are linked together into systems of communication and work processing within companies and where again the effect of employment is likely to be reasonably benign. The problems come with the third phase, when all these complex machines will be linked together into what he calls a 'telematic network' (1983: 171). In this move towards the boundaries of Science Fiction and Utopia, the effects on employment and human participation are more difficult to predict. The effect of the above on the occupational structure, the nature of the skills which will be required, and the general employment effects, are seen as complex. The author concludes the limit to the creation of jobs is the rate of growth in GNP and the share of the output that can be taken from production without impairing the productive capacity of the economy and social attitudes to income distribution.

The effects, however, may be unevenly shared, both socially and geographically, for structural economic change brings costs to the periphery when the older industries die. The decentralisation process may result from the inability to invest in new ways. As a key technological innovation, the microchip has not (as yet) stirred a new business cycle. Economics in turn has yet to come to terms with the new microprocessor technology; and its implementation has in turn neither unambiguously pushed economic theory one way or the other.

It is clear that firms are facing greater economic uncertainty and rising energy costs, with the attendant pressures on profits. The new technology offers labour-saving potential, and enhanced survival changes. However, the impact of new technological investment cannot be seen apart from the above general macroeconomic pressures.

Some North American evidence seems to suggest 'that automation is a trivial source of unemployment compared to swings in overall economic demand'. (Fallows 1984: 16). However, the auto-industry lost one in three of production workers between 1978 and 1982 for reasons that had nothing to do with automation – 'except, perhaps, that the auto-makers had been too slow to adopt it' (Fallows 1984: 16).

Given the loss of employment in the older 'rust-belt' industries, and job losses due to automation being introduced in existing plants, it compounds the problem to learn that firms based on making the micro-electronics technologies are not generating the compensating number of posts. If the market for new products incorporating microprocessors is growing, this may not help as the supply of such new capital – and consumer goods – may be satisfied by overseas firms. A Japanese analysis, on the other hand, is relatively optimistic from its own national standpoint because exports have sustained demand, but is cautious in the long term:

> Generally speaking, the development of new technology contributes to the economic growth and increased employment of a nation. However, ME machines reduce drastically employment at the work place where they are introduced. The effects of ME machines on employment tend to concentrate on regular male-workers and aged workers at the shop floor especially of the enterprises which make efforts to reduce personnel under the slow economic development.
>
> (Okubayashi 1984: 32–3).

However, in Japan, discharge and lay-off of workers has not previously occurred so often that it raises social problems. This is due to steady economic growth because of increasing exports, job security with transfer or retraining in big firms, and slow diffusion of new technology.

'De-industrialization' is, as one source puts it, 'the consequence of actions and inactions at the level of individual manufacturing units having a negative cumulative effect' (Winch 1983: 8). If the process is an evolutionary, and long-term one, it may accelerate further as the lack of technical expertise negatively feeds the supply-side, and as the lack of aggregate demand will call for less capital goods, which, in any case, will still incorporate productivity-enhancing and therefore labour-saving features.

The conceptual basis of discussion is unfortunately relatively weaker than the empirical evidence. Current employment levels have deep structural, cyclical as well as technological causes. The microchip did not in itself destroy the older industries. Longer-term questions of secular trends (the industrialisation of the Third World notwithstanding), as well as short-term questions of monetarist policy must be invoked. It is thus difficult to disentangle the various strands of the changes we are experiencing in the stagflation of the late 1980s.

Is full employment still possible in conditions of technological change? A sign of hope is that between 1965 and 1980, the US economy generated thirty million new jobs over this period of technical change, although much of this job creation was in the service sector. Even if British economy can compete internationally, full employment may be problematic. As a national survey concludes: 'when the wider effects are taken into account, the introduction of new technology offers the prospect of greater wealth and the possibility of increased employment opportunities in the economy as a whole' (Williams 1984: 210). Yet, if the new technology is introduced more slowly than in other industrialised countries, there will be a clear danger of job loss through weakening competitiveness without the benefits which technology can bring. But, if the economy adjusts to technological change, there will be considerable changes in employment, not just in terms of the loss of given types of job and the creation of new kinds, but also *vis-à-vis* the variety of skills which will be required for the 1990s.

Adam Smith's view was, perhaps, more optimistic when he claimed that: 'Each individual becomes more expert in his own peculiar branch, more work is done upon the whole, and the quantity of science is considerably increased by it' (Smith 1975: 10). He believed that the multiplication of the products of the different arts, as a result of the division of labour, which leads to, 'in a well-governed society, that universal opulence which extends itself to the lowest ranks of the people' (Smith 1975: 10). None the less, microelectronic technology will call for new kinds of expertise perhaps *of a different kind* from the increasingly specialised varieties Smith saw developing around him.

TECHNOLOGY, SKILLS AND TRAINING

Many writers take a determinist point of view, and argue that technology inevitably pushes organisations in a specific direction towards polarisation and de-skilling. A variant of this view is that the computer merely reinforces the *status quo* in organisations. Another perspective may say that the technology is neutral in its effect. Yet another school argues that it basically changes organisations and makes them more flexible. In any event, it is clear that technology has potential for the integration of systems and subsystems. This is probably why some label it 'new technology' and talk of a new 'industrial revolution', with varying degrees of clarity. They, in turn, anticipate that it will impose its logic on industrial society (see Lane 1988).

As far as enterprises are affected; the way computers affect the world of work has fascinated managers, computer experts and researchers for a long time. In earlier days of computing, some spectacular claims were made: enterprises would change dramatically, most managerial tasks would be automated, and organisations would become anonymous. Experience has taught us to be more cautious. Easy generalisations must be avoided. We must be cautious for it is facile to look at any overall impact of computerisation as there are so many different jobs computers can perform. The impact of computers on organisations should therefore be seen as problematic. Perhaps technology leads organisations in an unequivocal direction, and only reinforces conventional decision-making structures. Or we could argue that technology is neutral in its consequences. There is a fair degree of evidence however that it makes organisations more flexible, which is a view the present writers would basically endorse (see Campbell *et al.* 1989).

We can present technical non-determinism as a corner-stone of a realistic approach which does not exclude technical properties as having a consequence for work organisation, training, or skills, and other factors. But it does mean, however, as Sorge has pointed out

> that the socio-economic context impinges on the development, selection and application of technology just as much as the other way round. Technology is not autonomously given but developed to suit socio-economic purposes. Technical and socio-economic arrangements, including training and qualification structures, thus determine

> each other. Reciprocal determination then means that outcomes are ambiguous.
>
> (Sorge 1984: 23)

It is not possible to do justice to the complexity of linkages between the analytical boxes to be used in a study of technological change and industrial society, but a possible simplified model is set out in Figure 3.2. This schema attempts to introduce some of the relationships discussed thus far.

It is clear that given rapid technological change, craftsmen will need a broader range of skills in future. Boundaries between craftsmen's and technicians' work could blur with a need for more broadly trained engineers who are able to adopt a systems approach to implementing the new technology. Rather than the trend to specialisation of function continuing to a higher degree as was the case with industrialisation and technological change in the past, the application of microelectronics seems to imply the reverse. However, technology *per se* may not

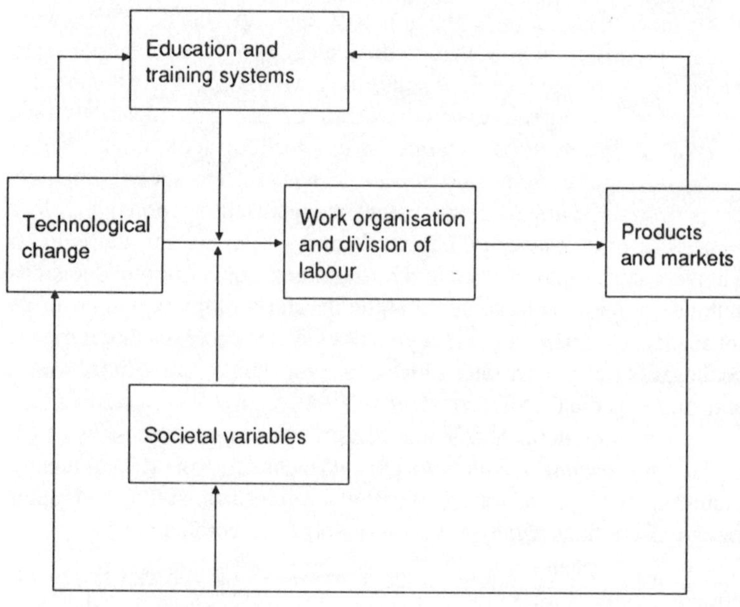

Figure 3.2 A model of technological change and work-related variables

determine the skill mix required. This latter factor will also depend on the management strategy undertaken, for example. The higher costs of 'co-ordination' in certain countries may in fact be related to the need to 'manage' higher ratios of relatively less widely trained (and hence less well trained) workers. We argue that the pivotal role of the factory as training ground can check the spread of over-specialisation.

If technology produces the need for new skills, the 'adaptation of training happens by very trade-specific introductions of information-technical and micro-electronics contents into existing training courses' (Sorge *et al.* 1983: 160). But it would be a mistake to just extrapolate existing trends, and we cannot assume that there will be convergence across industrial societies in terms of the above trade-specific training processes. Rather the effects will be both trade- and sector-specific as well as being 'societal' adaptations. As research we have undertaken previously suggests, *cultural/societal* factors shape what happens in specific national contexts. As has been pointed out:

> Interrelationships between phenomena within . . . analytically (separate) . . . blocks . . . bring forth nationally divergent manufacturing cultures extending across the full range of technologies. In particular . . . one can see differences in organization configurations arising because of the joint emergence of different work structuring and co-ordination, and qualification and career systems. It is primarily by way of the latter two systems that we see the societal effect taking place.
>
> (Maurice *et al.* 1980: 59ff)

The way technological change will mould future skills seems to be as follows: the wider the spread of new technology, the more technical information is diffused into a greater number of occupational tasks and occupations. The more this occurs, the greater its effect on personnel and organisational structures and therefore many other variables like unionisation, moderated by broad 'societal' contingencies.

Training systems developed at given levels of technological change may not be able to adapt quickly enough. Assistance may be on hand from a model in another sub-set, for example as some British firms now look to the German model, and as many others turn to the Japanese. It is however unlikely that models can be transplanted *en bloc*, nor should they be, as they grew up in particular circumstances. Determinist theories are misleading in so far as they imagine that technical change will always lead to similar training systems for desired levels of skills.

Because of the different and complex cultural/societal norms, training systems are veritable black boxes – much more so than most writers imagine. They are much more than the simple descriptions of institutions and certificates which are often conveyed in the literature. We therefore need to not only map the structure of training systems, but also the complex processes involved in their operation, as they operate over time. The evolution of training systems goes back into economic history, and even the more recent development is not satisfactorily recorded, let alone analysed.

The regulation of skill levels has taken place in a variety of ways, via the medieval guilds and later trade unions in some countries. Many critics see such regulation as outmoded, and favour centring training systems on the firm, whilst others invoke the state. It is a discussion which involves not only economic theories, but also social and political ones. Most critics agree that *ad hoc* arrangements are not enough. The pressures of international competitiveness dominate the debate. There are many interesting things to note concerning these linkages. Take, for example, technical change and training: it is clear that slow technical change may be linked with no great degree of training. Conversely, high rates of technological change involve greater inputs of training. It would not be difficult to look at national levels of technologically related investment and amounts spent on training at the macro-level. At the level of sectors and firms, there may be relationships seen in terms of money spent on each (as a first-step simple measure). Now, briefly looking at the relationship between training and skills, we can see how greater inputs of the former can lead to higher levels of the latter both at macro and micro levels. We have in fact talked about this link in a broad way in an earlier British–French–German comparative study (Maurice et al. 1980), with the German case as the most evident example of substantial training being found with high level of competence.

The relationship between technological change and skills is a more complex one, in so far as the latter acts as a constraint on the former. It is likely that high levels of technological change will be increasingly associated with hybrid (or mixed) skills where workers and managers will have less specialised training and broader ranges of taught capabilities to cope with the evolving technological challengers. Low levels of technological change may be found with low skill levels.

CHANGING PERSPECTIVES REGARDING TECHNOLOGICAL CHANGE AND CULTURE

Since the earlier Anglo-German project was carried out, many of the conclusions it arrived at have begun to be merged in a new consensus regarding technological change. Much of the academic writing on the subject prior to 1980 assumed some degree of universality regarding technology and organisation (Woodward 1965, for example), raising de-skilling to the point where it appeared as an integral and inevitable outcome of the economic system. Others (Child 1972, Clegg and Dunkerley, 1980) reacted against these more deterministic approaches, stressing the role of strategic choice in decisions over skills. This tradition did, however, imply an omniscience and omnipotence regarding management, to the extent that technology was perceived as having no effects other than those intended by the persons or groups implementing it. Wilkinson (1983) moderated this position by stressing the ways in which employees could influence the way in which technology was utilised. In recent years a more complex approach has developed, with writers such as Buchanan (1986) seeking to break the deadlock with a partial restoration of technology as an active agent – he defines it as a 'trigger variable' which enables different choices on the part of the groups involved, although it is seen to have no independent effects.

During the same period another debate merged over the validity of the universalist view. As noted earlier, Maurice *et al.* (1980) demonstrated how organisations varied in both shape and style between different national cultures – in this case, France, Great Britain and West Germany. This view informed the logic behind the study carried out by Sorge *et al.* (1983). Over the past few years, particularly since the appearance of Peters and Waterman (1982) there has been a considerable upsurge of interest in 'company culture' and 'organisational symbolism', the two concepts (ill-defined as they are) tending to mean the same thing although with a different emphasis ('symbolist' writers being less 'managerial' in their approach).

Exponents of the company cultural viewpoint do still concede that national or societal culture is the influence of the greatest consequence, with company culture merely superstructure erected on the surface. Company culture is interesting in other ways relevant to our study, however. Ray (1986) asserts (while disputing its likely effectiveness) that culture is being mobilised to fulfil a control function, filling a gap that has been opened up by a 'crisis of bureaucratic control'. We could maintain, in response, that such a crisis, if it is in reality a crisis, is only coming about because of an erosion of company culture through an increase in individualism, and that it was 'culture' which had maintained the appearance of bureaucratic control in the past. Hence an upsurge of interest in company culture, although, as Hofstede (1986) reminds us, an emphasis on the role of culture has been central to major contribution in organisational theory, including those of Barnard (1938) and Jacques (1951).

Among the reasons why such a 'crisis of culture' or of 'cultural control' should have occurred may be the nature of the technological changes that have been occurring, and the acceleration in the rate of change in which microelectronics has played its part. These technological changes themselves reflect changes in market, the 'crisis of mass production' referred to by Piore and Sabel (1985) which followed the energy crisis of the 1970s and the inroads made into Western trading economies by successive 'invasions' from the Far East. As a result of the crisis a number of trends have merged – a new emphasis on the small firm, on customisation, on flexible manufacturing, on niche marketing, small batches and networks of sub-contracting. Critics of this vision of 'flexible specialisation' emphasise how its success would be limited by large firms' reluctance to 'fade away like so many Cheshire Cats' (Williams *et al.* 1986). Others such as Shutt and Whittington (1987) have emphasised that much of the visible growth in small firms in recent years may be attributed to opportunistic 'hiving off' strategies by large firms, with the firms that result frequently being ephemeral creations. Further to this, one could add that exponents of flexible specialisation tend to refer to two distinct types of flexibility as if they were one and the same thing – meaning FMS and related developments (CAD/CAM etc.) on the one hand, and flexibility of employment and customer/supplier relations on the other. The conception, at least in the way it has come to be understood, tends towards Messianism and places undue emphasis on the role played by small firms, whilst idealising the forms of employment they imply. Sabel and Zeitlin (1985) offer a more complex view, demonstrating how different styles of organisation exist along a continuum (from authoritarian/feudal to pluralistic/democratic and so on) and how the form of organisation that emerges is determined by 'social struggles', of which the most important take place on a societal or national level. This stated position, for us, underscores the importance of international comparisons (see Lane 1988, Senker 1989).

We have no space here to enter into the different ways in which technological advances may lead to a 'crisis of culture'. One that immediately comes to mind, and which is clearly relevant to certain of the companies we visited, is the changing role of different layers of management and supervision. Since the 1950s, a debate has been continuing over whether technological changes involving different forms of computerisation will eliminate the roles of first-line supervisors and/or junior and middle levels of management (Leavitt and Whistler 1958, for example). Buchanan (1983) was surely right in asserting that new technologies may necessitate 'a reconstruction of management, particularly at lower levels of management' (see Campbell and Warner 1989a, 1990). The developments we have described in our case study of 'Rangefinders Ltd' (see Chapter 9) supports this view, as do, in different ways, the cases where project and matrix management had superseded line and staff hierarchies. These changes which have been taking place on a large scale may not necessarily remove some of the more fundamental problems that characterise British engineering. Commentators such as Hampden-Turner (1984) have described how the class system has eaten into engineering attitudes in Britain, so that engineers themselves divorce themselves as far as possible from any hint that they might be involved in producing anything. This is of course an exaggeration, although the overwhelming preoccupation with graduate recruitment that we encountered, as well as the national preoccupation with 'information technology' and hence 'pure' software engineering would fit comfortably within that framework.

At each company visited in the mid-1980s, (see Tables 3.1 and 3.2), we held extended and semi-structured interviews with one or more managers as circumstances demanded. The object was to assess:

1 How had product technology changed over the past five years;
2 How had process technology developed;
3 What had been the role of microelectronics in these changes;
4 What skill requirements had arisen as a result;
5 How did the company meet, or hope to meet these requirements.

Over sixty managers, and engineering managers with, in addition CAD/CAM managers, accountants, managing directors, general managers, project managers and systems managers were interviewed. Anonymity regarding both companies and personnel has been maintained throughout the twenty-four firms investigated.

Table 3.1 Companies/sites included in British sample

Site	*Code name*	*Main product range*	*Size*
S1	Circuits	Printed circuit plating lines	Small
S2	Packages	CAD hardware/software packages	"
S3	Devices	Measuring devices	"
S4	Testing	Microtesters	"
S5	Video	Closed-circuit and video	"
S6	Analysers	Chemical analysers	"
S7	Systems	Conference systems	"
M1	Materials	Material testing equipment	Medium
M2	Office	Office automation systems	"
M3	Battle	Battlefield communications	"
M4	Micro	Microscope, semi-conductor testing equipment	"
M5	Telco	Telecommunications	"
M6	Controls	Control systems	"
M7	Stabilisers	Stabilisers, control systems	"
L1	Communications	Controls, communications	Large
L2	Aero	Aeronautical engineering	"
L3	Mechanical	Mechanical engineering	"
L4	Avionics	Avionics	"
L5	Engineering	Mechanical engineering	"
L6	Computers	Computers	"
L7	Measuring	Measuring systems	"
X1	Aviation	Aviation equipment	"
X2	Alarms	Alarms devices	"
X3	Rangefinders	Range-finders	"

Note: Number of employees: S (Small) = up to 200; M (Medium) = 200–1,000; L (Large) = over 1,000

Table 3.2 Status of British companies/sites studied

Site	*Single-site company*	*Site of British parent company*	*Profit centre of larger concern*	*Site of nationalised company*	*Site of American-based company*
S1	X	–	–	–	–
S2	X	–	–	–	–
S3	–	X	X	–	–
S4	–	–	X	–	X
S5	X	–	–	–	–
S6	X	–	–	–	–
S7	X	–	–	–	–
M1	–	–	X	–	X
M2	–	–	X	–	X
M3	X	–	–	–	–
M4	X	–	–	–	–
M5	X	–	–	–	–
M6	–	X	X	X	X
M7	–	X	X	–	–
L1	–	X	X	–	–
L2	–	X	–	–	–
L3	X	–	–	–	–
L4	–	X	X	–	–
L5	–	X	–	X	–
L6	–	–	X	–	X
L7	–	–	X	–	X
Aviation	–	X (Group)	–	–	–
Alarms	–	X (Group)	X	–	–
Range-finders	–	X (Group)	–	–	–

Chapter 4

Summary of findings

RECRUITMENT VERSUS TRAINING

As we review the data from the present British study, the most striking aspect of the findings is the overriding concern with graduate engineering skills. Although a number of (usually smaller) firms reported serious shortages of technicians and skilled prototype workers, almost all shortages in the larger firms were shortages of graduates, usually electronics hardware or software, and perhaps most importantly, shortages of 'firmware' and systems engineers.

Whilst such shortages (particularly those in the last-named category) are undoubtedly important, and are likely to be predominant in Germany also, we might perhaps question the centrality of these shortages to the skills and training debate as evident in the Butcher Report (DTI 1984). The Butcher Report takes a rather simple 'pump-priming' approach to training in the economy. Interviews with DTI officials seemed to suggest that the general view was that training was the responsibility of industry, and the government could merely provide the overall strategic skills (high technology graduates) that industry could not be expected to provide. For the rest, the traditional British approach of 'on-the-job' training (not to be dismissed, since it could be claimed, as it was by several managers, that formal training is less efficient and only serves to create the framework whereby learning can occur through on-the-job experience) would make most of the running for the rest.

To an extent, the DTI was merely reflecting the emphases of company HRM, recruitment and training policies. The second part of the Butcher Report, intended to deal with skills at technician level (the shop-floor or intermediate levels were not covered at all), could only work with generalities, since companies appeared to have very little precise information on the numbers and qualities of the skills either possessed or required at this level. At the graduate level, on the contrary, the most precise figures can be obtained almost instantly. Immediately, it comes to mind that the graduate shortage is given attention because it is bureaucratically the most comprehensible and the most accessible. Furthermore, the emphasis on the need for high technology graduates plays a part in the pressure on universities to concentrate on vocational courses – the much publicised 'switch' (in 1985) of funds from arts and social science to new technology subjects meant a drop in the total number of students, since more money was required per student in the latter subjects. This emphasis on the universities ignores the fact that countries such as Japan have, if anything, a less vocational university syllabus, the emphasis on vocational training taking place elsewhere.

Overall, large British companies appear to prefer recruitment to training. Skills are bought off the shelf wherever possible, the simplest (though in the long run one of the most difficult) means of doing this being to recruit 'ready-made' graduates. Smaller companies, and in particular companies with fewer than fifty employees placed a greater emphasis on training and development of staff, with a high premium on versatility made necessary because of the need to overlap with such small numbers. In these companies one often found people with ONC or HNC level qualifications doing work which in larger companies would be seen as wholly the preserve of graduates. Interestingly, it was the smaller companies which made proportionately greater use of local educational establishments. Often, they could only recruit school-leavers by offering them day-release at a local college for an HNC. On receipt of the HNC after four or five years, these trainees would invariably leave to join a larger company which was able to offer them a career structure in line with their increased expectations. The only advantage small firms had was that they offered the possibility of more varied and interesting work at each level of skill than the large companies.

In these firms, the diminishing returns on graduate skill were a problem. Although an effective 'half-life' of five years was seen as an inevitable result of a high technology industry where skills were quickly superseded, the problem may in part be due to large organisations themselves where initiative tends to become submerged in the large concentrations of graduate specialists.

Even if it is more a question of the speed of technological change, the problem is exacerbated by the emphasis on recruitment of new graduates rather than the training or the retraining of the old. The Group Training Manager of a large electronics company strongly repudiated the 'Butcher approach' in this respect, and was organising a substantial retraining drive to redevelop the existing personnel in the company. Interestingly he was also instrumental in keeping the apprenticeship and technician training programmes at full strength, with some mixing of mechanical and electronic disciplines, despite the relative decline of craft and technical labour in the production process.

WOMEN IN ENGINEERING

Another change that is likely to occur regarding the composition of engineers, is the recruitment of more women. In one case (company L4), an engineering manager was of the view that merely increasing the number of engineering places at universities was not going to solve the problem of shortages since it sometimes meant more second-rate engineers. Believing that most of the 'natural' engineers had been recruited from the male population he was trying to influence women's career choices from secondary school level, before they were likely to follow more conventional career paths. Companies such as Case Ml and Rangefinders, were encouraging women to join but apparently only limited numbers were as yet applying. Half the new graduate recruits at Company L7 were women, although as yet they tended to be graduates in computing rather than in electronics hardware.

Managers commenting on this trend imply that women are less interested in the 'hard' side of engineering. Research findings (Keenan and Newton 1986) suggest the contrary. Where women have managed to penetrate into the engineeering preserve they have been found to have less difficulty adapting to the work in the business setting. Unlike their male colleagues, they are less likely to be 'prima donnas' regarding expectations for the job. Keenan and Newton find that the mismatch on expectations is a cultural phenomenon, and has little to do with the university training, contrary to our managers' perceptions. However, these perceptions may cause role conflicts for women engineers (Bailyn 1987).

Both the above findings have considerable relevance regarding the problems the companies in our study have encountered. The recruitment of women engineers would present itself as a solution for the impasse that often exists between cost-conscious managers and engineers with negative views of business and commercial applications.

GRADUATES OR TECHNICIANS

Technicians had two types of advantage over graduates regarding flexibility. On the one hand, as already stated, they were more inclined to see beyond the bounds of the specialism; on the other, they had lower expectations regarding the technical interest of their work and were thus able to grow and shrink with the job as it went through different stages in its level of technical complexity. Graduates, on the contrary, were trained to over-specialise and were likely to encounter problems when a long-term project 'died on them'. Their skills were more likely to be under-utilised if they were unable to find an appropriate niche in the company. In some respects a problem may be said to exist whereby companies need skills which, in many cases, can only be conferred by university training, which in its turn creates expectations which companies are not always equipped to meet. Universities are geared to the incremental advance of 'science' as a 'public good' (in economic terms) rather than the more contingent applied approach implicit in the German concept of 'Technik'.

With complex product technologies both approaches are of course needed, many companies resolving the problem by offering two career paths to engineers, one general-managerial (leadership of projects etc.), the other purely technical (internal consultants etc.). In many, the university-trained technical specialists were seen as aliens to the extent that many found it difficult to adapt to company cultures. At one company they were described as 'the sort of people you lock away and throw a banana to occasionally to keep them happy'. Another consciously divided its recruitment between 'boffins' (who were to be kept to the minimum) and the 'generalists' who would be recruited with a view to advance within management.

In an attempt to find the necessary equilibrium beween the two extremes, some companies (notably Case M4), had developed a complex 'matrix' pattern whereby seven project managers (chosen entirely on non-technical grounds) supervised 130 projects. Project groups were multidisciplinary and involved a total of 545 engineers. Another fifty constituted a 'pool' of differing levels of expertise and experience which would be assigned to a variety of projects as and when required. In this way, very few people worked on only one project, groups were small enough to enable knowledge to be passed from one discipline to another, status was acquired outside of the project leadership structure as well as within, and specialists in the pool were moved around enough to keep their 'applied' knowledge of the products within reach of their technical knowledge. The same pool enabled new graduates to acquire general experience without the initial and sometimes limiting experience of being assigned to one narrow project. We may mention that the nature of current technical product development in all of the higher technology companies meant much fluctuation in design work groups. Company S6 had academic connections stretching back over a century, as did Case M4, over a shorter period. Both companies had begun selling scientific instruments to the universities – in the case of Company S6 they recruited direct from their customers – and had branched out into the commercial market as education spending was cut. They retained structures from that environment – high individual initiative, minimal supervision, informal co-operation – which appeared to be well suited to the needs of their product technology.

In some other companies, where design and development had expanded in a company with low-skill traditions (Case M3) there was something of a cultural rift between the highly trained laboratory staff and the rest of the organisation. One manager spoke of how the Ph.Ds 'exercised authority through their knowledge, not through their personality', something at odds with company traditions. Interestingly, this was the only company where the personnel department readily admitted that they had problems with a de-skilled workforce (in this case women circuit-assemblers) who complained of the increased boredom of their work and their alienation from the final product. No 'pendulum effect' appeared to be at work in this case of skill polarisation. Elsewhere (Rangefinders Ltd), one manager believed that due to increased standardisation of circuit designs and software in the near future, the 'balance of skills' would move away from electronic engineering towards the 'packaging' skills of mechanical engineering and systems assembly, which would increase in importance in value-added terms reversing this trend.

Clearly, for a large number of jobs involving a high level of technical understanding, graduates will be essential. However, for many others, it may be more feasible to 'train up' technicians than managements are prepared to accept. The question of unionisation may be relevant here, along with the trend for graduate technical management to insist on hiring graduates even for jobs which waste their abilities. Graduates frequently have higher expectations than companies can fulfil, and their disaffection can be costly and may only be recuperated through expensive retraining, if at all. Technicians, on the other hand, were held to be more flexible and able to adapt to the demands of each task. In this respect, apprenticeship training possesses advantages over university training. Among the larger companies few, apart from Aviation Ltd, were making systematic attempts to make changes in this direction. Even there it remains to be seen whether the strategy for relieving high-level skill shortages through comprehensive training up from the craft level would be politically acceptable within the company.

BLUE COLLAR SKILLS

No company apart from the very smallest (Cases S1, S4, S5 and S7), reported any shortage of blue-collar staff at all, although large companies such as Case L1 said that such shortages came in unpredictable waves – three years previously it had been impossible to find a prototype wireman or woman in that part of South-East England. Shortages of this and similar types of skilled worker are regional rather than absolute in character (unlike shortages of graduate specialists). Skilled engineering labour abounds in the older industrial regions but is more scarce in the South and West where many of the companies were situated and where the local unemployed have not been trained in the appropriate skills. House-price differentials between the regions make the transfer of labour more problematic, as recent initiatives to import skilled labour from Teeside into South Bucks have shown. Such differentials affect experienced professional engineers also, as case M5 demonstrated – the company no longer attempted to recruit beyond the limits of the Thames Valley. Smaller firms in the South-East are more severely hit (notably case S7) since they do not have the resources that some large firms are able to commit to importing workers during the week and returning them at weekends, an unsatisfactory system which may be likened to having internal 'gastarbeiten'.

Another important development is the polarisation of skills between technicians doing non-routine and routine computer-assisted work. Mixtures of operator and maintenance skills are also needed at sub-technician level, to part-programme CNC machine-tools for instance, as well as coping with faults and breakdowns. Mechanical and electrical/electronics skills in future will need to be combined. However, where apprenticeships have been maintained, there has been a clear shift from craft to technical and from mechanical to electronics.

Elsewhere, shop-floor skills were being marginalised by their exclusion from the design process. With CAD/CAM/CAE, the emphasis has moved to centralised databases facilitating accurate repeats, modifications and customisations. This change fits ill with traditional craft practices. One manager in Company M6 stated 'Everything must be right before it leaves the "drawing board". We've got far too many people down there who'll correct or adjust something and then won't tell us they've done it – so the next time it'll come out with the same fault again.' Whether the shop-floor's secrecy over this arose from lack of understanding, or whether it represented a discreet form of bargaining we cannot say.

As far as craft skills are concerned, criticisms have been made of the 'workerless factory' view of automation (Sorge *et al.* 1983, Senker and Beesley 1986). The need for skilled workers who understand the process and who can intervene at the point of production is being recognised in some cases. In this case senior managers had become aware that 'right first time'(the slogan behind CAD/CAM, CIM, etc.) meant 'right fifth time' if they were lucky. Although greater improvements had been made in terms of precision and accuracy, the communications of information in digital form left plenty of room for errors. Being involved in flexible small-batch production of a complex order, Rangefinders, for example, found it essential to retain shop-floor skills well beyond those apparently required with CAD/CAM. The problem in this case was how to give new machine operators the experience that the older ones had acquired in pre-CAD/CAM days, and how to reassure the operators about their future when they had seen all their programming responsibilities shifted between the drawing office and the machine shop appeared to work as a zero-sum with no clear solution that could be acceptable to both parties and to management.

CUSTOMISATION

Customisation reflects the increasing tendency for firms to be (or at least profess themselves to be, since such concepts become fashionable) market-led, which implies a form of flexible niche-marketing rather than reliance on a few standard product lines. Differentiation in slow-growth markets means that many firms will be required to spend more effort in satisfying individual customers, although always looking for how the extra development on a customised product may be utilised in larger batches for a broader market. A significant number of the companies studied were geared to servicing one large customer already, usually the Ministry of Defence, although the policies regarding competitive tendering had latterly undermined the security of their position, even before the run-down of defence orders in 1990/1.

Customisation has made a particular impact in the high technology commercial sector where a degree of stagnation (compared to the initial high growth) has coincided with increased customer awareness, resulting in firms having to pay more attention to a customer's specific needs. Case S3 for example, had not only to customise its systems for its clients, it had also to install them, train their staff and oversee the implementation over a long period. Close association with such customers did however pay extra benefits – it provided a source for recruiting experienced personnel, as well as guaranteeing future business. In the event, company S3 went bankrupt as a result of banks' refusal to bail the company out of its cash-flow problems, despite a significant number of orders outstanding to the company. Interestingly, the sale price of the company did not take account of its user base which constituted its main asset, built up by customisation.

If their ability to provide personal service and a high level of customisation constitutes a major strength of small firms, one of their most serious weaknesses is their reliance on sub-contractors for whom the small company represents an insignificant proportion of their business, particularly if the sub-contractor or supplier is a large established firm. With lead-times increasingly a major factor in competition, production-planning difficulties through late delivery can be fatal, as can a large customer's tardiness in paying.

There is a link between the trend towards the customised product and the much-publicised skill shortage (or mismatch, it might more accurately be termed). Sluggish growth helps engender greater competitiveness (on quality and lead-times rather than just price) and market differentiation as flexible niche marketing is applied to make up for the contraction of the mass market. The customer is in a position to issue more specific demands. In the first place, as said above, the customer demands higher levels of quality and reliability, as well as prompter delivery. It is often stated that firms compete now on quality rather than price. This is not strictly true; what most of the companies reported, with surprising regularity, was that the customer wanted higher levels of quality from increasingly flexible and sophisticated products, and that they wanted this at the same price as before. Price is thus by no means secondary. What is of course clear, is that in the sectors investigated there was no market for inferior goods whatever the price. It would be wrong to deduce from that that the firms do not take the opportunities to 'rationalise' and cut costs just as before. Almost without exception, the firms covered were either shedding labour (usually unskilled) or at the very least increasing the turnover of the company without a corresponding increase in the number of jobs (what one might term the hidden job loss of technological advance).

Customisation leads to more sophisticated technology (to meet a wider range of specific needs). A 'knock-on' effect works as one firm's customers will themselves be suppliers of another and so on. The demand for more capable and reliable technologies works backwards through the system. It does not only move backwards, although the movement is generally seen as demand-led. It also works forwards from the key technical innovations themselves. The advent of the microchip led to a veritable explosion in customer expectations. Once the same capability as before could be encased in a far smaller space, customers demanded that the remaining 'unused space' be filled in terms of more elaborate functions and higher quality levels, and wanted the improved version for the same price as the old – in real terms.

EMPHASIS ON DESIGN AND THE NEED FOR HYBRID SKILLS

The product/market nexus, whereby products have to be customised for customers whose expectations have radically increased as a result of faster technological change, appears to lead to a much greater emphasis on design and development as a source of profit, and a consequently greater emphasis on skills in that area. This is not to say that the problem may be solved merely by producing more of the technical specialists of whom there are certainly too few. The direction that technical product development has taken has led to the generating of more and more complex systems, where the contributions of the different disciplines are integrated into more and more complex relationships. Specialists are unable to cope with the convergence that occurs, which is aggravated by the need to modify products for different customers. Furthermore, with each innovation that 'facilitates' this process, the possibilities – and consequently the expectations – increase accordingly and the problem becomes more serious.

The introduction of CAE, enabling repeats, may in practice mean that modifications and adjustments to designs must be more precise and that 'muddling through' is no longer acceptable. That is, that even as an improved capability for 'repeats' helps engineers to offer customers shorter lead-times, it reduces the lead-time horizon in which they are bound to operate if they are not to lose out to competition. It may therefore be that if the benefits of integrative design technologies are to be realised, then engineering skills will have to follow the same path of integration. One CAD/CAM manager interviewed envied the Japanese phenomenon of the so-called 'flexible engineer'. As with most Japanese phenomena, this was seen as arising from cultural traits absent from the West. The manager concerned expressed the hope that the technical integration embodied in CAD/CAM and CAE would by-pass western cultural trends towards fragmentation and specialisation and necessitate the development of more hybrid engineers. The need expressed by respondents in cases M4, L4 and L5 in particular, for systems engineers and people qualified to take an integrative and 'product-wide' view confirms the need if not the means of fulfilling it.

The universities have responded to these developments through a greater emphasis on 'firmware' (software and hardward combined). Courses have also burgeoned in answer to the demand for more specific hybrid subjects – microprocessor engineering for example, where the emphasis is not on the mechanics of the chip itself, but merely on where it should be placed. In line with our general picture of trends and counter-trends, there seem to be two contrary movements in this area. On the one hand, there is a need for people who know, for example, electronics hardware and software thoroughly up to a high level in both, thus making them generalists in the 'polymathic' sense. At the same time there seems to be a need for a more 'dilletante' style of generalist, who knows less of the ingredients of the different disciplines but more about how they relate to each other. In some companies engineers are concerned not with basic design, but with choosing suitable sub-assemblies out of catalogues in line with customer requirements, a trend which, in the long run, could favour what we have termed the dilletante generalist. This move returns us briefly to the role to be played by technicians. In M1, we were told how it often transpires that design engineers, through lack of practical, as opposed to theoretical experience, overlook the effect of the positioning of cables, for example – oversights that technicians would not usually make. A combination thus suggests itself, whereby generalist engineers would provide an overview in detail by technicians. It appears already to be the role in some companies regarding the division of labour between design laboratory engineers and CAD draughtsmen, between which groups the arrangement of tasks may be said to have always contained a degree of arbitrariness.

Companies have always employed hybrid engineers in different forms – aeronautics is an example of a well-established hybrid discipline. Because of the publicity surrounding 'information technology' many companies involved in areas such as aerospace began to employ software engineers as a priority. In practice, they found that such recruits had a low level of product awareness, or did not 'know one end of the product from the other'. Such companies now appear to prefer either firmware engineers or hybrid engineers (aeronautics or otherwise, depending on the sector) who have been taught software. In some cases (L2 for example), the company will teach them itself. This preference for 'their own' engineers can only be encouraged by the fact that electronics hardware and software engineers now have access to a much wider range of sectors of the economy than their predecessors had. Aerospace or mechanical engineering companies cannot compete with the salaries and conditions being offered by companies in the City of London for example. One unfortunate aspect of the 'information technology' vogue, however, is that bright university entrants will tend to gravitate towards specifically electronics-related fields, assuming that that is the only way to guarantee employment. As a result, courses in fields such as aeronautics, we were told, tend to be undersubscribed.

SIMPLICITY AND COMPLEXITY IN CYCLES IN PRODUCTS

A further area where customisation and related trends produces paradoxical development is in the nature of the product itself. The process that occurs there, is, of course, related to our comments concerning generalist and specialist skills. We were told that, with the arrival of microelectronics in the late 1970s, products became simple as components and sub-assemblies were compressed and miniaturised. What occurred then was that customers expected more functions at the same price as before, and the extra product 'space' was filled out. As the new configuration itself grew too complex, it too was simplified, and so on. The overall trend was for the parts to become simpler and the whole to become more complex with each 'twist'. This results in the situation described above, where hybrid skills geared to integrating different parts of a system begin to become more important than the specialist skills geared to concentrating on the parts. The way the cycle operates is set out in Figure 4.1 (Campbell and Warner 1987a).

THE FUTURE OF ENGINEERING SKILLS

Management strategies have a tendency to focus on and contain key constraints on their decision-making power (Campbell 1985). In the past, in Britain at least, dependence on shop-floor skills – consequently

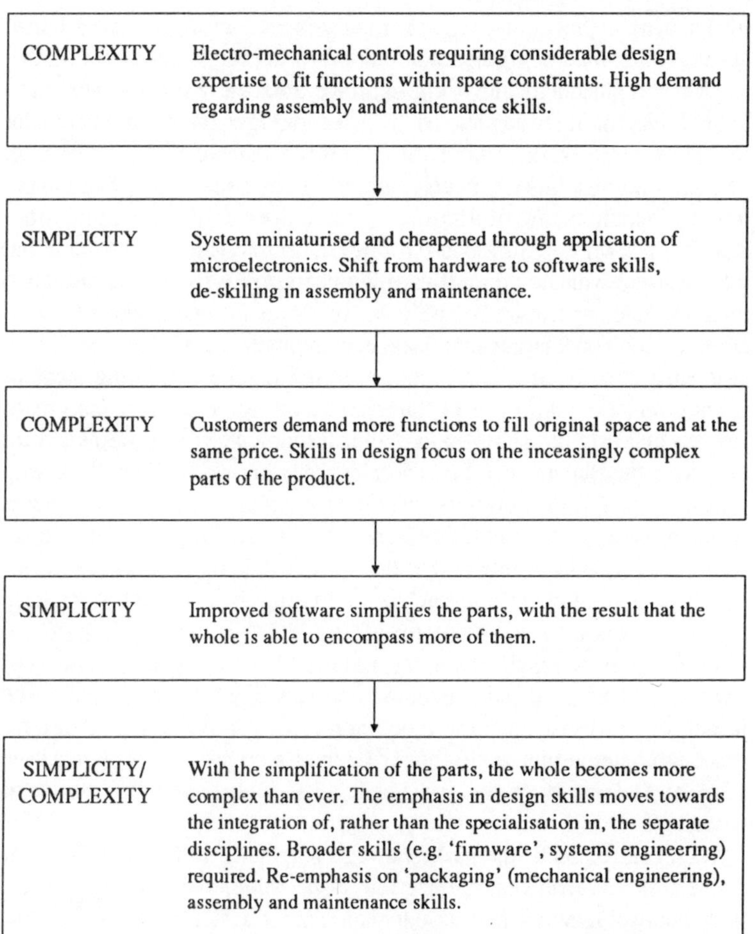

Figure 4.1 The simplification/complexity cycle in microelectronics product applications and possible skill implications

Source: Campbell and Warner 1987a

on organised labour – was seen as such a constraint, technology usually being applied to minimise it. With technical advance labour costs have fallen as a proportion of the total sum (though regular 'shake-outs' continue to suggest that the shedding of labour is seen as a prime means of gaining competitive advantage). With unemployment and considerable weakening of trade-union power, the 'labour constraint' has been replaced by the 'skill shortage'. As we have already said, unemployment is accompanied by a serious geographical mismatch regarding craft skills. That is one side of the lacuna; the other is a genuine and absolute shortage of engineers and scientists of graduate level. The shortage of these, we have found, is restricting the possibilities open to all but the largest, most prestigious and most technically advanced companies in engineering. Whilst compensatory measures such as conversion training, up-skilling of technicians (which could be done on a larger scale) and extending the useful working life of graduates already in the company through regular retraining (which again could be done on a far larger scale) may be applied, one may suspect that, as with shop-floor labour, technology will be deployed to eliminate the dependence on scarce engineering skills (Campbell and Warner 1986a). No firm in the sample gave voice to any such intention although companies such as M2 and M4 were attempting to exercise tighter budgetary control on the work of designers. Most were developing both their approaches to finding new recruits (through stronger links with educational institutions, including some customisation of teaching courses) and endeavouring to reduce any excessive turnover of such people once they had been recruited. One large firm (L4) saw no sense in looking for ways to rationalise design and development, since the 'pool of expertise' they had gathered was seen as the main source of both value-added and competitive advantage.

As we have seen, the skill shortage perpetuates the dominance of larger firms in electronics, since few of the smaller firms can provide both interesting work and an adequate career structure for ambitious recruits. In the purely software-based area this is not so true, since many software graduates appear to be drawn by the higher rewards and more congenial conditions of small software houses. For some people, a large bureaucratic company may be anathema, and they may look to hold relatively higher positions. Usually higher pay was given as the reason for leaving, although the training manager at company L1 emphasises that money was merely the 'socially acceptable' reason for going. Dissatisfaction with work and with the organisation was often more important. The moves being made by a number of companies towards closer monitoring of engineering costs, while understandable, may worsen this situation in the short term, emphasising as they do the different expectations of graduate engineers and their employers.

As we have already stated, the interest inherent on the task itself provides one avenue for reconciling the two. Case L4 managers, for example, claimed to lose very few to small software firms and cited the case of one of their best software engineers who left his work on flight-control systems to earn higher pay designing video-games. After six months, he returned out of sheer boredom.

THE 'SUSPECT' SOFTWARE ENGINEER

On the whole companies seemed to view software engineers with suspicion. Company M1 revealed how when software engineers were brought in, more traditional engineers began to follow them in demanding improved pay and conditions. In a long-established mechanical engineering firm (Case L3), it was described how the arrival of new categories of 'whiz-kids' with higher qualifications and in departments of rapidly increasing importance, caused some rocking of the paternalist boat as older employees resented the way they were falling behind in relation to the newcomers. In the process the organisation grew larger, more formalised and began to distance itself from its traditional relationship with its employees. A similar process was being effected at Company M6 with moves to differentiate rewards through a more systematic appraisal scheme, enabling engineers of higher ability, and more up-to-date knowledge, to be paid their market value (in practice the level of engineers' salaries is decided in broad terms by meetings between the major electronics employers in which the parent company of this medium-sized site took part). The multi-nationals, Cases L6 and L7, which prided themselves on a high-quality intake and minimal turnover, tended to pay above the national market rate. Older firms, in particular those whose origins lay outside the electronics field (M3), were disinclined to pay extra and were thus more likely to sponsor people (in some cases their own staff) through higher education. This is becoming the trend in Japan, where such initiatives are intended to forestall a massive shortage of engineers and technicians over the period up to the year 2000 and possibly beyond.

For those who possess electronics hardware or software qualifications, it is therefore a seller's market. Increasingly, there appears to be a trend whereby the company's obligations to such privileged employees overshadow the reverse (something which is beginning to be true with general managers as their turnover increases through an erosion of traditional 'company loyalty'). In Case L5 for example, new graduates were wary of an induction process that would leave them with company-specific skills, although the large and well-established company was of the type that, in the past, would normally have employed such people throughout their working lives. To attract people in, the company had to provide induction training of eighteen months (rather than the three months that were considered necessary) so recruits would be eligible for institution membership. Once the course was completed, 25–30 per cent of the trainees left the company. Then, turnover would fall below 5 per cent in this area and the company would be concerned that they had lost all their best people. Although problematic, as this example shows, training in the future may be a major means of attracting people into companies, including regular retraining and updating after several years, since, even if some companies were prepared to let employees' skills go to waste once their best has been given in the first ten years, the employees' self-interest may force them to do otherwise, or they may stay away in the first place.

The strong market position of these skills may however contain the seeds of its own downfall. In software particularly, companies are finding that, in view of the shortage, they can make do with fewer. Case L4, having found fewer software engineers than they needed, adopted the approach of having one software engineer operating as consultant to four or five hardware engineers, who had received a measure of conversion training and who needed help only on the toughest problems. Had they been available, the company would have had six software engineers as a group. Elsewhere, apparently tired of software engineers who 'didn't know one end of the product from the other', a company (Case M6) instituted a practice whereby engineers in the electronics domain must work in both hardware and software for five years before they are allowed to specialise. Courses have been arranged with local colleges so that over two years engineers from hardware can learn software and vice versa. Some are sent on specially devised systems-engineering courses.

As well as the switch to 'firmware' and a greater emphasis on conversion training, another threat to the software career comes from off-the-shelf and reusable software. Companies L4 and L5 both suggested that the need to generate software from scratch was diminishing and with it, ultimately, the desperate requirement for software specialists.

ELECTRONICS ENGINEERS – BUILT-IN OBSOLESCENCE?

Although electronics hardware specialists are in a stronger long-term position in the industry in that it is easier for them to convert to software than vice-versa, they may be under threat from similar developments, which, although logical extensions of current technical trends, are likely to be favoured by companies that have found recruiting in this area troublesome. Cost-consciousness and a desire for integration has led to a large number of companies investing heavily in CAD, CAD/CAM. The first economies from these innovations are derived from the production area, with fewer repeats being necessary (although implementation, effective operation and evaluation of these sytems is a highly complicated process the aims of which are not always fully understood at all management levels). With CAE, the potential is there to make greater use of 'captured' designs (Senker 1984, 1990) which is seen as essential for effective customising and recustomising of product designs. CAE potentially lends itself to closer supervision of the progress and cost of projects, if linked to a company database. The managing director of one company ('Rangefinders' Ltd,) recently declared to his managers that in his view it should no longer be necessary to design products from scratch. The result must only be some falling-off in the numbers of electronics design engineers and a greater emphasis on more 'routine' adaptive work in the drawing office or the labs to which technicians displaced from obsolete production control departments have been moved.

Some managers, as we have already stated, believed technicians were easier people to employ in that they were more flexible than graduates. Whereas graduates' skills declined if their work became routine, technicians appeared to be able to shrink or stretch as the nature of the work demanded. Hunt (1984) described the 'skill twist' that resulted from the introduction of robotics – jobs-displaced findings confirm this, but leave open the possibility of a further twist in the opposite direction. At Rangefinders Ltd, one manager stated that, although the company would still be hiring new electronics engineers for the next five years at least, the time would soon come when through CAE on the one hand and new forms of circuits (such as application specific integrated circuits – ASICs), electronics design would become a more routine affair and the emphasis would move to the 'packaging' no-redundancy policy. What is clear from this is the way senior managers (usually of a mechanical background) regret the extent to which their discipline has been relegated to the periphery. That technical advance presents the possibility of making the focus of operations move to the 'periphery' is clear from what we have said concerning the simplicity–complexity cycle in products.

Evidence of such a shift towards mechanical and mixed engineering skills came from Case M4, who said that for their purposes the type of person most in demand and most difficult to find was a good mechanical engineer with experience of working with the semi-conductor industry.

In the next three chapters, we intend to explore in greater detail how these broad trends manifested themselves in the three categories of firms investigated, namely large, medium-sized and small sites. A case study approach will be adopted in each instance, followed by a set of conclusions and implications for HRM strategy, subsequently in chapters 8 and 9 respectively.

Chapter 5

Case studies: large sites

L1: 'COMMUNICATIONS LTD'

General

The first large company case (L1) presents a concise picture of the changes occurring in the electronics industry. The site was occupied by three distinct product divisions – Traffic Controls, Industrial Controls and Data Communications – all of which shared the same site facilities. In recent years the product divisions had come to overshadow the functional organisation over which they were superimposed. Their relative strengths regarding personnel, resources and markets had also changed.

In recent years, the growth in Traffic Controls' business has been offset by a relative decline in that of Industrial Controls (sensors for the nuclear industry for example). This division and its products approximates most to a traditional mass-production operation. The third division, Data Communications, has in recent years grown to equal the financial turnovers of the other two divisions combined. Concerned with packet switching, networking and network interworking, Data Communications is organised on a project basis and is increasingly concerned with specialised unit production for specific customers. The rise of Data Communications has thus seen the rise of the Project Management Department of the division in relation to functional departments over the site as a whole. Regarding the shared resource of manufacturing, project managers in particular (or their parent department) were seen as having significantly increased their power. With single projects sometimes running up to £25m over four years, or 10 per cent of site turnover, the amount of financial risk on a single unit assembly on the shop-floor is greater than can be entrusted to the decision-making of the manufacturing manager alone, it seems.

Furthermore, with such projects, design and production overlap to a considerable extent, blurring traditional boundaries. The floor space occupied by manufacturing had been almost halved in five years, reflecting changes in the relative turnover covered by sophisticated products where design is the key variable and components and sub-assemblies are increasingly bought in. At L1 the peculiarity of the structure was physically evident on the shop-floor where a £50m single unit was being assembled only six feet away from a line producing 25,000 miniature radiation detectors a year.

Since 1980, product emphasis in the company has moved from being product-oriented to being system-oriented. This has had major effects regarding skill requirements, and regarding the relative strength of divisions and departments.

Batch-size and 'information potential' are important variables here. Industrial Controls, like the other division, uses microprocessor-based designs. However, for their large batches of sensors, optical character recognition devices printed circuit board (PCB) inspection systems, information potential is relatively low and the software load is comparatively limited. Although microprocessors had increased in importance, the emphasis was still on non-electronics disciplines such as physics. Industrial controls had increased from forty to fifty engineers over five years.

Traffic Controls represented a traditional strength of the company, the first such products having been launched by them in 1932. Within the division there had been a major shift in the direction of microelectronics over five years, with computer-control replacing electro-mechanical control in the product areas mentioned above. The growth in the number of engineers (from fifty to ninety) was accompanied by an emphasis on software skills, and involved retraining as well as recruitment. The speed and complexity of the data transmission now required meant a more significant shift to 'information technology' than in Industrial Controls, and a greater emphasis on design than on manufacturing.

Data Communications are involved with telex switching, packet switching and network interworking. The division has benefited from the very large growth in demand for fast communications between computers and, increasingly, between networks. With miniaturisation and increasingly flexible hardware, there has been an upsurge in software requirements. This has been reflected in the composition of the design and development workforce which has gone from 50 to 220 engineers in five years, with 90 per cent being computer science graduates. The emphasis of Data Communications has moved away from the more standardised of its products (telex switches etc.) and towards highly customised networks and network management services, with a corresponding trend towards the major products already referred to. Data Communications' rise has been facilitated by the installation of CAD on a large scale, for PCB hardware design, approval, and test program simulation.

Personnel

The company employed 1,187 people, consisting of:

Engineers	400
Indirect engineering staff	35
Other indirect production	140
Direct production	200
Field Services	200
Site Services	50
Marketing/Commercial	70
Finance	50
Personnel/Training	7
Senior management	35

Over five years, the number of graduate engineers had doubled and the number of direct production and administrative white-collar workers had halved. The reduction of manual employment by 400 during this period had apparently passed without incident, being largely carried through natural wastage. Reassurance over job security, given competitiveness, was seen as the main task of industrial relations staff regarding the manual and craft unions at the plant. In return for redundancies, contract work had been phased out for the most part.

Whilst the jobs lost were mostly in the unskilled category, the period saw significant trends regarding the polarisation of skills. In Field Services, for example, traditional craft skills were in retreat. More reliable equipment and the use of microprocessors rather than relays has meant fewer skills are required at depots. At the same time the increased capital risk (up to £10,000 for a PCB) has meant a greater reliance on service engineering skills on the company premises to which faulty components are sent for analysis and repair. However, as the microprocessor-based system grows more complex over time, a broader understanding is required from maintenance works. Small teams of highly skilled workers/technicians were expected to take the place of craft maintenance in the future.

Polarisation had occurred in what remained of the direct production area. The very decline of manufacturing compared with design in the competitive strategy of the company actually protected the skill-level in the former area. With greater design-intensity, the size of batches produced had fallen sharply, while the complexity and variability of operations had increased, rendering them unsuitable for automation. To the extent that such automation would be possible it was unlikely to occur, since technological investment was largely going into the expanding design domain.

This development left management more dependent on shop-floor manufacturing skills than they would perhaps have liked. Skills involved in the assembly of highly complex PCBs had usurped the position of wiring-related skills which had declined due to microprocessors. Hybrid techniques had become important – with flow-soldering of PCBs (a skilled operation in itself), half the components had to be inserted after the others had been soldered. Also, while wiring individual components was not as challenging as traditional wiring, the complexity of products meant that operatives needed to be conversant with a wider range of components and methods than before. Management seemed to be particularly dependent on trouble-shooting craft workers/technicians who were able to deal with the whole range of problems. To an extent then, the increased complexity and flexibility of the product technology has necessitated flexible and complex skills on the shop-floor. The composition of the workforce had changed in so far as the switch from electro-mechanical to PCB assembly meant an increase in the production of women workers.

Skills and training

As we have described, graduates now made up almost half the total employed, most of them working in design and development engineering, with a high proportion purely in software. The orientation of the company towards both design and software intensity led to skill shortages becoming a major preoccupation of Personnel, in the way that industrial relations might have been ten years earlier. At present, such shortages are almost entirely confined to the graduate software sector. In previous years, there had been periodical shortages of skilled manual workers such as prototype wirepersons. These had, however, reflected the comparatively recent arrival of such industries to the area and are a system of mismatches in the national labour market. With software the shortage has, of course, been more enduring. The company was competing for a share of a small and specialised cake, against all the most prestigious electronics companies (and many in other sectors). They were also forced to compete with small companies and software houses, where salaries are up to 25 per cent higher (as well as less paperwork), more congenial working conditions, and sometimes more advanced and up-to-date equipment, often made them more attractive to recruits. Thus, while the company had no current vacancies for technicians, draughtsmen, test engineers, etc., they had twenty-five software engineering vacancies. Recruiting was not the end of the problem – keeping recruits was often even more difficult. The size and prestige of the parent company could perhaps ensure a steady stream of new graduates (who made up one-fifth of the graduates recruited). It was after their recruitment that the attraction of small companies was a danger. New graduates were expected to stay for only two to four years unless they received promotion. However, as the Training Department report, graduates were likely to undergo two crises within this period: one after only six to eight months, and another after between fifteen months and two years. For this reason special technical training or management development courses were made to coincide with these vulnerable periods. This policy was an implicit recognition of the fact that whilst higher salaries is the most common and most socially acceptable reason given for leaving, other reasons, such as social environment and the nature of work, are more significant.

Shortages are a problem and are likely to become more of one the more the company's product strategy is rooted in its software design capability. As we have seen, the software component of Data Communications products was increasing, as was the proportion of company business taken up by Data Communications, in line with market trends. Shortages of software specialists may thus act as a constraint on the company strategy. It has not been unknown for the company to lose a contract on account of staff turnover in this area, and there was little doubt that contracts which the company was suited for in other respects were not pursued if they entailed the hiring of a significant number of extra software engineers. Contracting work out was one solution, although it ran the risk of quality or delivery failures. Contracting extra personnel in was against the policy of the company and kept to a minimum (six out of 400 engineers were on contracts). None the less, the rapid growth in the number of Data Communications software engineers reflects a surprisingly high rate of success in recruitment. Interestingly, the neighbouring site, which housed part of another company in the same group and which consisted solely of 400 engineers (production was carried out on a site 200 miles away), had started a practice whereby technicians were trained up to become software engineers. Technicians were said not only to be up to the right standard, but tended to be far less likely to leave the company, compared with university-trained software specialists.

The training function played a conscious part in alleviating labour turnover problems, in the process of which the training officer acquired a role akin to personnel trouble-shooter and counsellor. The sub-department (halved from eight to four during the company's retrenchment) was thus involved in the socialisation of recruits into the company. Although the vast majority of training was purely technical, and geared to dealing with the problem of the low 'half-life' of information technology graduates (four and a half years compared with a desired figure of five and a half) through updating, courses were also provided to help graduate scientists adopt a more profit-oriented approach through problem-solving exercises. The training officer held that graduates had a problem in adjusting from university-trained 'pure-science' to industrial applied science – 'it's not the best solution we want but the right one'. With increased customisation, involving the addition or deletion of technical functions in a product, such understanding was crucial – 'after all, engineering is the art of making for £1 what any bloody fool could make for £5'.

A major training programme was set in motion in 1983 when the potential of staff turnover to disrupt contracts was recognised. It was still considered to be embryonic and was expected to expand in the future. With a budget of £400,000 (less than 1 per cent of turnover), it was not large, although to it could be added the cost of having individuals away from their jobs for the duration of training. The costs of not training were seen as much higher, if not amenable to measurement. Also, internal training under the tuition of company specialists was not covered by this figure. Such training was also given to non-graduates. The training officer had conducted a basic electronics course for operators (and potential operators) on the site, with he and another instructor learning at nights to keep one step ahead of trainees. This type of training was amenable to being used as a means of selection for the job being trained for. With graduates, use was being made of personality tests (about which the company were not sceptical) in conjunction with training to assess candidates for promotion. This was one of a number of cases where a company was using training as the instrument of a hidden agenda.

To some extent it was envisaged that in the future, training would win back some of the ground from pure recruitment which had been a failure to the extent that turnover was high in areas of short supply. Up to ten local trainees were being taken on each year for training over a six to seven year period. If they were treated properly, it was expected that, being local, they would show more 'inertia' and be less inclined to leave. Recruits of A-level standard could have a larger role, given the proper training, since with them, unlike graduates, 'nothing had yet been cast in stone'.

Such a move towards lower formal scientific qualifications was seen as consistent with the development of technology; thirty years previously a Ph.D. would have been required to do what would now be an ordinary programmer's job. Hybrid skills were likely to arise as the skill emphasis in design moved further from the technical intricacies of individual components. Ten years previously, the company had sought to fill the gap regarding generalist skills (still in short supply) by recruiting systems engineers. Now, the trend was more towards a microprocessor engineer who would know nothing about how to design an integrated circuit, but would be expert in positioning microprocessors within design parameters. This would be the logical conclusion of the company's strategy whereby the printed circuit board was the main source of value added in the product.

L2: 'AERO LTD'

General

Case L2 was a major aeronautical engineering company. Traditionally, such a site would see the design and construction of an aircraft from start to finish. In recent years, with an emphasis on multi-national projects, such sites are beginning to specialise far more, establishing themselves as 'centres of excellence' regarding one or more aspects of aircraft manufacture. These changes have been accompanied by investment in new technology.

This is necessary in design and production and testing, not only on account of the increased sophistication of products and the resulting complexity of safety considerations, but also because the sites involved in such projects are so far removed from each other. This means there is less room for trial and error at the assembly stage. Where this assembly takes place, and which sites supply which components, is as much a matter of political lobbying as of 'business rationality'.

Personnel

In recent years there had been a limited shift from blue- to white-collar workers, so that the latter now outnumber the former. In 1980 there were 2,500 blue-collar employees and 2,000 white-collar employees. By the mid-1980s, the total has increased by roughly 500 with over 3,000 white-collar and 2,000 blue-collar workers. The annual employee turnover was 6 per cent, rising to much higher levels the more sophisticated the skills possessed by staff. Among blue-collar workers, most of whom lived locally and had always worked for the company, turnover was very low.

No one on site was employed on a contract basis. This may have been connected to the comprehensive vetting procedures which are required by security. Labour was often traded between different plants of the enterprise, on a reciprocal basis. Some functions or jobs were often put out to sub-contractors, notably in the area of technical drawing.

Skills and training

Aeronautical engineers and software engineers were in short supply. The problem with the former was that existing aeronautics places at universities were not attracting sufficient numbers of applicants. A priority thus existed to inform schools' careers services that jobs were available in this field. Clearly, the message had not been effectively communicated in the past.

As far as software engineers were concerned, similar problems had occurred to those in other firms. One engineer described how, in his view, even a mediocre specialist in this field could become 'a big fish in a small pond' if he/she moved to a software house. People often achieved status there who would never have achieved distinction in a firm of this size. This problem manifests itself both before (in the sense that they never apply) or after software engineers take up employment at the company.

As far as avionics were concerned, what was wanted in the long run was not software engineers but systems engineers who had expertise in a specific area such as aeronautics, hydro-mechanics, etc., and who also had computer ability. Some moves in this direction (creating more 'all-round' specialists) had been made through computer-adjustment courses for such engineers, arranged with a regional university.

It appeared that the avionics department had been able to cope with shortages by spreading the work out among those engineers that were available, and had managed to expand relatively swiftly. Personnel planning presented difficulties – in an industry organised very much on a project basis with much complex dealing between sites no such comprehensive plan seemed practical.

Training

At the time, there were 240 trainees, divided equally between white and blue collar. In-house and external training covered between them around 700 people (300 outside, 400 inside). Of these around 150 were blue collar.

Most of this group would not be learning microelectronics-related skills. A high proportion of workers overall (around 2,000) came into contact with microelectronics in one form or another. Out of 200 workers in the machine shop, 60–70 operated CNC machine tools. Of these twenty-seven had, apparently on their own initiative, attended evening classes in programming, and were thus able to programme the machines they used. Of 150 production electricians, around 50 per cent had microelectronics training.

A wide variety of training courses were utilised by employees of the company. In some respects, the training officer was a liaison officer between the various agencies and institutions involved (in-house training tended to be departmentally based). Overall, it seems fair to say that the company's policy was to keep training to the minimum necessary. In the Personnel Manager's opinion, 'over-training' led to unnecessary dissatisfaction. Training was thus directly geared to the use to which skills would immediately be put – a result of financial constraints as much as anything else. Basic craft training, on lines approved by the Engineering Industry Training Board (fitters etc., formed around 700–800 of the 2,000 total involved in production). Apart from this basic training little actual instruction was carried out on the site. Otherwise 'arrangements were made for people to go to experts' for training. If FMS became a reality in coming years (as was the plan) then in-house training might become more significant and far-reaching. Less people were now leaving after completing their training (this had presumably posed a problem in the past). The new projects the company was involved in after a period of slack orders were seen to have increased interest in staying with the company.

All manner of institutions were used for training purposes and all levels of staff were involved to various extents. Some emphasis was placed on staff development, as might be expected where technology, business strategies and structures are relatively fluid.

New technology and skills

The adoption of CAD had led to a skill reorientation among planning engineers. Rather than the merging of design and production that was observed at L1, for instance, CAD had led to a partial by-passing of both by planning staff. The latter (held to be more versatile and better geared to change than production engineers), had undergone retraining in order to appreciate the possibilities of the technology. Both the drawing board and the circuit diagram are done away with, making whole systems 'captured' in the information base the criteria for choice. In the end existing plans are matched against the particular contingencies and constraints involved in the project (regarding wiring etc.) This was intended to cut out production control in the near future, with the final process being a chain of directly linked CAD, CAP and FMS, leaving the operator merely administering the workload. The main constraint was said to be capital financing for such techniques – the necessary training or retraining of personnel being a function of the investment available.

A key role was played by the avionics department. Avionics is a hybrid discipline that grew out of the fusion of airborne electronics and systems engineering. Microelectronics, whether in the product or through the use of computers, has had a major effect on the avionics department. Five years before only two or three computer terminals were in use. The engineers are trained, partly outside, partly on site (by outsiders) and generally reach an acceptable level of competence in computer skills after twelve weeks (although they are expected to improve these substantially when on the job). At the same time, the hydro-mechanical side of systems had reduced in size as the actual systems used move towards avionics. Avionics soon covered twenty-five of the seventy systems posts, the twenty-five being divided between electronics engineers (12) and others (aeronautics), software (13). Problems were encountered in finding the appropriate personnel for this expansion of avionics (which has doubled in three years). Recruitment has been halted after the completion by an earlier project, so that most engineers were over 35. Two of these (from the mechanical side) volunteered for retraining to fill the new posts. Otherwise it was necessary to start from scratch. The preference for these posts, as far as the avionics department head was concerned, was for graduates either in electronics, aeronautical or aircraft systems engineering, then for systems designers, and only then for computer graduates.

The latter were seen as being inclined towards the narrow view – they would not have the knowledge of the product necessary for giving due consideration to questions of safety, for example.

An understanding of how each part could affect the whole and ability to design systems within narrow constraints was seen as central to avionics skill. These developments were being accompanied by an unprecedented expansion of computerised testing. Testing technicians now carry out far less mechanical operations and use different tools, such as signal generators and logic analysers. This entails some expense – the latter cost £16,000 each. As is the case with hydro-mechanical engineers, some knowledge of microelectronics is now vital in the training of test engineers or technicians. The latter formed the white collar 'shop floor' of the avionics section. The avionics design team was held responsible for testing, although this responsibility did not extend to customers, once the product is sold. Most airlines would have their own elaborate testing departments. Over 800 persons were involved in testing on the site.

L3: 'MECHANICAL LTD'

General

L3 was a long-established paternalist firm involved in aeronautical and mechanical engineering (both middle-batch and unit production). Over the past five years it had accelerated its move into new product areas so that its traditional speciality now accounted for only 15 per cent of annual turnover. These changes had important implications for both the company's culture and skill structure.

Personnel

Very significant shifts had occurred both in the number, composition and qualifications of company employees. Since 1980, there has been a 30 per cent increase in the total (about thirty-five). A further 300 new personnel had been recruited, partly offset by the retirement of 100 older employees. The trend was towards a younger, more qualified workforce.

The breakdown of the workforce was as follows:

Management	180
Clerical/Administrative	200
Skilled Workers, Technicians	590
Specialists	190
Un/Semi-skilled	300
	1,360

Overall the number of skilled workers and technicians increased by around 40 per cent, while the number of unskilled workers has declined by around 20 per cent.

There were about twenty contract workers at any one time. These were usually in the area of electro-mechanical design or draughting.

Skills and training

Few problems were encountered with skilled workers – many of these, such as fitters and electricians, were taken on as they were shed by other large companies in the region. Problems higher up the scale were not so much to do with the numbers of graduates available but on the type of qualifications. There were too few aeronautics engineers, and other engineers tended to be too specialised. This was seen as part of a trend in education that was to be resisted. The advantage of aeronautics engineers was that they would have knowledge in a wide range of spheres, of which mechanical or electrical engineering would represent only one part, whereas other engineers would be qualified in just one. This need for more broadly qualified engineers reflected the changes that had occurred in the company's business. The company had started out operating only in the mechanical and aeronautics areas. In recent years, it had expanded into production engineering (use of CNC and robotics), and electrical and electronic engineering, including software. The latter has only very recently begun to play a part in the company.

The company embarked on, and completed, the first major phase of a programme of up-skilling. This was a development of some significance, taking place as it did in a firm of well-established paternalist traditions. These were represented by the fact that the managing director was the son of the original founder. As it happens, the managing director was keen on innovation and the professionalisation that went with it, and therefore was prepared to make all the necessary investment.

This fact enabled the company to expand into new areas – CNC,

CAD, FMS – and to pay market rates for suitably qualified staff. Unlike Battle Communications Ltd, in many ways a similar company, Mechanical Ltd were prepared to upset the 'apple cart' of their more senior employees by not giving full priority to the maintenance of traditional differentials. The job of the personnel manager had been radically affected as a result.

The changed skill-profile reflects a change in the company's product strategy. In 1980, the traditional product line accounted for over half the annual turnover. By the late 1980s, it counted for only one-sixth. One of the rapidly growing areas which has superseded it (in relative terms), is that of tow-targets.

These are, in terms of value, 75 per cent microelectronics and software. They are relatively cheap due to the use of FMS for small-batch production. Much of the electronic components and sub-assemblies were, however, still bought in, and it seemed likely that the company would remain rooted in mechanical engineering and the complexity of its newer products will be reflected in a focus on design, with their manufacturing contracted out. The company is becoming increasingly involved in larger-scale projects, with a number of different firms. The largest of these concerns the design and construction of a remotely piloted vehicle for reconnaissance. Most of the manufacturing is subcontracted. This is essential in that it is impossible for a firm of L3's size to acquire and effectively use the required expertise in all the areas involved. In this case there was a need for knowledge of composite materials. This was obtained through the purchase of a company which specialised in this area, for which the learning curve is purportedly seven years.

Avoiding over-specialisation may be costly in the sense that qualified generalists tend to demand more money. One of the mistakes L3 was attempting to avoid was that made by Westland, who had acquired on-site expertise in every relevant area and manufactured as many of the components as possible themselves. They also carried out costly customisation of their products. The result was a large number of organisations grouped on the same site, which proved unmanageable. The project manager wanted, in view of this, to sub-contract as much as possible. The advantage of this is the fixed price and the avoidance of having to pay a high hourly rate to large numbers of workers who may not be occupied for much of the time. At one time each project would have staff of each type assigned to them specially. Now, more of a matrix structure prevailed, so that resources, in the area of design for example, are shared more evenly.

The shift in company composition and identity had its repercussions in the area of industrial relations. In the past, the company had no formal bargaining structure and no trade union recognition. This remains the case, but with the recruitment of many new personnel and the slow collapse of paternalism (the company began to implement single status), something began, informally, to move towards a more adversarial system. Consultation such as it was, was carried out through various committees. What had happened was that a significant number of fitters and electricians had been recruited from Plessey where they had been members of the AUEW and the EETPU. As a result, the pay review recommendation had been rejected for the first time. Senior management, who had never even had to deal with an unofficial strike and thus had no bargaining experience, caved in immediately and added an extra week's pay across the board.

L4: 'AVIONICS LTD'

General

Case L4 was a major site within a nation-wide group, which in turn forms part of a major British engineering company. They were concerned with a wide range of electronics-based products for the aerospace and defence-related industries. The site was divided into divisions which serve as profit centres (our enquiry was carried out through one of these), concerned with control systems, as well as through the centralised training function.

Since microelectronics were first applied in the mid-1970s, and most particularly since 1979, almost all the components involved in L4 products have been miniaturised. Although this temporarily relieved the design constraints posed by the limited space available in airborne or seaborne vehicles, the extra 'space' has been filled with increasingly diverse and complex functions. All this has increased design input, as opposed to production which has been simplified and automated (with a 'concealed' rather than an actual job loss). Quality, reliability and safety of design were together the major focus for competitiveness, although the need to expand resources in design meant economising in production costs. The company were operating in a more sharply competitive climate than before, the reason being the competitive tendering policy of the Ministry of Defence (MOD) on the one hand, and the concentration of the civil aviation industry on the other.

Personnel

There were 5,500 people employed on the site as a whole. Of these, 380 were employed in the Controls Division. Of these, in turn 150 had degrees or the equivalent. Detailed figures are not available beyond this breakdown.

Skills and training

Training was co-ordinated across the site as a whole and was directed by the central training function in line with the requirements of divisional engineering managers. It was noticeable that of the two managers interviewed the divisional engineering manager enjoyed higher status than the deputy training manager. Training was still seen as a junior service, however importantly it was now figuring in the company's competitive strategy.

Graduate recruitment

The sponsorship scheme for electronics undergraduates was in its fourth year. Around thirty students join it each year. Overall, the site takes in about 100 graduates each year. This was expected to rise to 130 or 150. In general, it is their first job after university and they stay for around three years. On the whole, the company claimed to have little trouble in keeping people in the long run, even in the case of software engineers. The reason for this (in so far as it was true) was that the work offered by the company was of an interest that would be hard to come by elsewhere, except in a few other large companies. Any software expert who left to take a less intellectually challenging job for more pay, soon came back. Outside defence and aerospace, no truly challenging jobs in the field were seen to exist. This is not to say that competition for the best brains, against other companies in the same field, has not reached a new intensity in recent years, paralleling the increased competition due to competitive tendering by the Ministry of Defence (MOD). Just as, on the one hand, the latter requires greater attention to cost-effectiveness and efficiency (including getting designs 'right first time') so the former requires better training and promotional opportunities for those the company wants to attract. The two areas are not unconnected in that greater cost-effectiveness means the more efficient use of better engineers, while also making large numbers of them less necessary in theory 'cutting down the amount of design insurance time'. However, the effects of the drive for greater cost-effectiveness were more likely to be felt in the lower grades and the production sphere. The latter had not grown with increased turnover, although there had been no redundancies. Engineering, on the other hand, had grown. Asked whether at some stage they might not move to rationalise the engineering sections as design–production links became more streamlined, the managers said this was unlikely. The human resources in engineering design and assurance were seen as a key part of the company's competitive edge regarding quality and were not to be subjected too strictly to crude measures of cost effectiveness.

There was always the possibility that a hidden job loss might occur as it had in production, in that more efficient techniques and organisation might render large scale engineering recruitment loss necessary, even if the division stayed at around the same size. The engineering manager, while seeing reputation for quality as being an important selling point, also recognised that people often did not pay a great deal more for a better product. Clearly the balance between the concentration of expertise and the meeting of budgetary targets was a difficult one to quantify, just as in product development the decision as to how far one proceeded with refinements and modifications was more a marketing decision than a cost accountant's, although the frame of reference of each might move closer together under pressure.

Skill shortages

When skill shortages occur, the initial reaction is to sub-contract. This was being done with respect to the five software engineering vacancies that remained unfilled. On the other hand, they had found through organisational changes that they needed fewer than might at first have been supposed. They had found, for example, that rather than recruiting, say, five software engineers, they would recruit one and train half a dozen of the design engineers to carry out their own software tasks, only bringing in the software engineer when absolutely necessary. In this way, the shortage of specialists might be seen as a stimulus to develop wider skills amongst existing employees. Again, we come across the recruitment versus training dichotomy. We may also note the possible future of software engineers in such companies as being that of specialists carrying out a service to more flexible general design engineers.

Another way round this type of shortage is to technically 'iron out' the need for repeated specialist human intervention. This can be done by orienting the division towards reusable software that does not need to be verified every time it is used. In this respect the software specialist, much of whose value consists in software assurance ability, may suffer the same fate as the testing technician has suffered through the introduction of computerised testing and the decline of diagnostic due to the cheapness of chips and PCBs.

As elsewhere the specialist shortage may be encouraging a step away from insistence of degree-level qualifications. There are moves being made towards the deployment of 'software technicians' on a large scale, with software engineers as such existing only as the co-ordinating nucleus of experts.

Women engineers and the skill shortage

In L4, managers did not, for example, see the increased number of university engineering places as a solution to the skill problem. Skill shortages are a question of more than simply the number of people available with the right qualifications. In many cases, the increased number of places and the general awareness of the security and future of engineering as a vocation simply meant more mediocre engineers. Recruiting managers were turning increasingly towards the first-class engineers who might be lying dormant among the female population from whom, very probably all the 'natural' engineers had been recruited and it was a question of diminishing returns. Firm L4 attempted to get in early by approaching girls' schools in the area in order to spread awareness of the possibilities of engineering as a career for women, since the choices against engineering are made well before university entrance is attempted. They had received, so they said, no help from senior mistresses who were seen as still believing that only genteel subjects such as English were good enough for their pupils. Thus, no headway had yet been made. Out of 150 engineers, the division had four women engineers who had been recruited over the past two years.

Management training

Courses were designed and structured by the company and given in-house engineering candidates. These were sent on them in order to prepare them for their next promotion, which invariably takes them further away from their technical knowledge base.

Arrangements did apparently exist with outside providers of course to have their courses customised. First line supervisors also attended in-house management training courses in technical supervision, designed by the parent company Staff College in conjunction with other institutions. The most favoured course was an Information Systems M.Sc.

A third area of training was the graduate conversion course, the company being praised for its work in this area by MSF in its first skills and training bulletin. Under this scheme, graduates in arts (notably Philosophy) and Music were taught software. This coincided with the policy of the company not to take on computer science graduates where it was possible to avoid them. At present, computer science graduates filled ten out of fifteen of the software engineering posts at the division (vacancies notwithstanding). Conversion schemes of the type mentioned above seemed to be still at the earliest stages unlike at a related organisation (where according to one source turnover is appalling, with around 60 per cent of conversion graduates leaving in the first year). This was believed to be more a reflection of the company culture than the greater rewards to be had elsewhere – a view shared by the training manager at L1 who regarded the usual diagnosis, that software people just want more money, as hiding a multitude of sins. Conversion courses of this type are residential and last three months.

In practice, therefore, most conversion graduates get the most significant part of their training on the job, having only learnt the basics on their conversion courses. They do, however, represent an attempt to break into a new recruitment market, presumably also with the hope that the people thus taken on will be less 'computer-motivated' and will thus have a more general approach to the company's product. With the inclusion of more software training in electronic engineering courses, there is a possibility that this kind of conversion course may be less important in the future than might otherwise have been supposed, particularly since most engineering managers will stress how their preference is for 'good hardware engineers who can do software as well'.

Conversion courses at different colleges are also offered to software engineers who wish to acquire hardware skills. These involve full-time attendance for the first or second six months of a given year to produce a hybrid software technician. Also, on a larger scale this time, an eighteen-month sandwich course was embarked on to convert maths graduates into software engineers (maths graduates were considered to be of a higher calibre than their computer science counterparts). Of twenty who had been sponsored previously, only one had left the company. This was seen as reflecting the company's skill in selecting those who were likely to develop the greatest commitment to the company. This scheme reflects the company's current strategy of both committing more resources to training and at the same time paying premium salary rates. It refers us back to our earlier discussion on the twin strategy of capital and human resource investment at the specialist level. The training department organised the up-skilling of computer programmers to software engineers via a 3–4 year HND sandwich programme based at a local college, for which a software staff lecturer had been provided by the company. Interestingly from the hybrid skills perspective, electronics and physics were both included on the course syllabus. We may also note in passing how being the largest employer in the county may confer an advantage for a company when liaising with local educational institutions eager to justify their survival in economic terms. Unfortunately no figures were available for those attending the course.

A further example of closer links with educational institutions were those being forged with a regional polytechnic. These would apparently involve some kind of trading in expertise, specialists from the company giving their services on courses which would at the same time be adapted to the needs of L4 and similar employers.

Another area of training was that carried out by the company as a service to its customers. The company ran training courses on a commercial basis for airline service technicians, courses for which they provided full-time staff. These would deal with the installation and maintenance on the control equipment produced by the company in which microelectronics has come to play a dominant role. In this case, the training is an extension of customer service. Before the advent of microelectronics, customer service was fairly limited due to the technical capabilities of the airlines. However, the speed of technological advance has made many of these skills obsolete, hence the need for training by suppliers.

Like sales staff in more mainstream commercial companies, it is not only the customers' personnel who are trained with the advent of each new product, but also the L4 trainers themselves. The higher the level of technical sophistication in products, the greater the emphasis on customer training as part of the sales package and the greater the emphasis on the training of both salesmen and customer trainers themselves. This training is not the preserve of the site training function, but that of an Aviation Service and Repair Division.

Retraining was provided for the technically displaced. Due to natural wastage, this was not seen as the problem it might have been. As far as older engineers were concerned, some in-house retraining was provided, though not on any significant scale. Engineers were expected to use their own initiative in keeping abreast of developments in their field – for this reason all the relevant periodicals were made available.

Testing staff had been trained on a one-to-one basis as to how to use automatic test equipment. This was apparently done by project-team engineers. In general, technical training of this type was carried out on the job. In the recently expanded area of CAD, conversion courses had been given for those draughtsmen who were to use it. Whilst those who were displaced (around ten in number), were given retraining for technician-type tasks. The division in question only possessed three people who could strictly be termed technicians. There were more in some other divisions, notably the Instruments Division which also had a larger number of precision mechanical engineers. There had once been a number of these in the Controls Division – now there were very few. Technicians, like draughtspeople, tended to be found in the facilities shared by divisions. This was a reflection of the policy of maximum utilisation, both of capital equipment and of personnel below the engineer level, whose numbers were proportionately smaller as a result. In the field of testing, it is significant that software testing remained a divisional function – it is also significant that, as said before, software testing was an area which the division was hoping to make economies in in the future. In general, the technician/craft apprentice route seemed to be on the decline to say the least, being increasingly confined by technological advance to the divisions where electro-mechanical engineering still survived. Employment in the technician and skilled worker categories had also suffered through the trend towards building fewer prototypes in the division itself.

Training for new equipment, in design production or testing was always carried out formally once it had been bought even though in practice most would be learnt on the job. To save money, the company would not usually send the operators away on a course at the suppliers or wherever, but would save on the accommodation costs by having a lecturer brought in. Even where this was not possible they would always see to it that the 'hands on' part of the training took place in-house. One exception had been the training of secretaries in word processing. When office automation had been introduced, all those concerned were sent away to be trained. This appears to have been necessary only once – once people know how to use them, any newcomer could learn on the job through the knowledge of those present. This is presumably the pattern regarding the new equipment generally.

As far as production and assembly are concerned, courses had been provided at the site training school. Wiremen and women had received training and retraining in electronic assembly there. Production engineers had also been participating in courses at the school as new techniques (which there would be planning for) were brought in. These techniques are dealt with below.

Future developments

Two further developments regarding recruitment and training needs are worthy of attention.

1 Specifically with the increasing specialisation and customisation of technology towards which the division is moving, management foresee the need to develop a pure electronics design facility to produce those devices that large-scale producers, like Motorola, would not bother with. This will necessitate the recruitment of highly specialised silicon-device engineers in the near future, who tend to be scarce and command a high market price. It is in line with the trend towards more specific wishes by customers who want more functions in their sales package.
2 Over the coming years greater and greater amounts of specialist knowledge in general were expected to be required in the company products. This may imply a greater concentration of skills in the company either to displace skills at the customers' sites or to fill in for those which the customer may not be able to acquire. It is also a reflection of the ever-increasing ambitions of customers regarding the inclusion of sophisticated functions in products.

At the same time there will be paradoxical development in the nature of the product itself, which will have an effect on the skills required by the supplier. Modules will become more standardised and more expertise will be oriented towards the development and deployment of the whole as a system. Moves are also under way to expand the sub-contracting of projects either to or from the company. The company is already selling its expertise to multi-company projects, and this is likely to represent a larger proportion of turnover. Thus, some trends point towards the operation of such large companies primarily as pools of expertise, with most of the production process taking place elsewhere. The idea of sub-contracting whenever it is economic to do so seems to have taken root.

To save on the massive recruitment costs that might be involved in developing a large enough body of expertise on the site to cope with these strategies, the company is looking towards the development, from within its own ranks, of a creature known as the Flexible Engineer so that it can accomplish a wider range of tasks with the same personnel. This, if properly implemented through broadening of skills through training, might mitigate the effects the skill shortage has had on the company. These, like those of automation on the employment levels of skilled and semi-skilled workers, have largely been hidden but have been equally serious. People with a given skill have been stretched because of the lack of people with similar skills to back them up. As a result there has been too little slack in the system to embark on too many new projects or to experiment. Thus, the effects of the skill shortage can be measured – not in the number of projects that were abandoned (this is very rare in this area of industry) but in the projects that were never started or even thought of.

L5: 'ENGINEERING LTD'

General

Case L5 was a very large, complex mechanical engineering site. The increased sophistication of the systems it produced entails the co-ordination of literally 100 distinct disciplines, many of them hybrid. Integrating the work of these represented the rationale (and a major challenge) for the CAD/CAM investment project, one of the very largest in the country, then undertaken by the company.

We have chosen to concentrate on one division on the site. This division, Technology Support, was the one most concerned with microelectronics involvement in the product. In practice, this means quality-assurance of work carried out by sub-contractors and suppliers. The company had yet to come to a final decision as to whether it wants to include a full-blown microelectronics design/production facility in its organisation, although the collapse of major sub-contractors concerned with one project in the 1970s did precipitate it into carrying out its own electronics work. To carry out similar work on contemporary projects would, given the complexity of technology involved, necessitate a substantial investment but remains a 'piece of unfinished business' as far as the Governing Board were concerned.

Technology Support operates very much as an organisation within an organisation – it has its own administration, carries out its own recruitment and planning. One somewhat paradoxical result of skill shortages, however, is that it occasionally has to borrow people from suppliers in order to check suppliers' work – although presumably not from the same suppliers. The process is not sub-contracting in the usual sense in that the whole of a project, including the design, is put out, with the firm only playing an active part once the work has been completed.

Two means of control, largely unconnected, are used over the sub-contractors – on the one hand the L5 finance department checks the cost estimate put in by the supplier, and on the other Technology Support check its quality and applicability to the rest of the L5 system which is put together on site. It is perhaps curious that cost and design/quality are checked by completely separate departments.

Personnel

The breakdown of personnel in the division was as follows; figures for 1970 are in brackets. The reason for taking 1970 as the cut-off point is simple – the aerospace industry went into digital equipment, and later microelectronics, far earlier than the rest of industry.

Analogue electronics applications/ design engineers (the numbers of applications has increased by a far greater proportion, but has been largely in the digital area)	35 (30)
Hydro-mechanical engineers	35 (15)
Electronic Design	12 (9)
Controls performance	15 (0)
Simulation	2 (4)
Software assurance	6 (–)
Total	105 (58)

Although the software assurance group appears to represent a radical change of emphasis, it should be pointed out that, although the group as a formal unit was only set up in 1985, the task was being carried out before that date by disparate people. It is also worth noting that the hydro-mechanical side had grown significantly, whereas one might have expected it to lag behind. Not only does it represent the importance of this hybrid discipline in the particular product of the company, it also emphasises that however important electronics or microelectronics becomes, it is the fit between the technologies as a whole which determines the quality of the end-product. According to the director there was a tension between specialist and generalist priorities in the work of the division – while a high degree of accuracy was necessary in separate disciplines it was the whole system that was important, something which many individuals found hard to adjust to. He said, 'People tell me how good a component of sub-assembly is – but I'm only concerned with what the whole product is going to be like.' There is a contrast between the two newer departments or groups within the division. In software assurance (the most easily sub-contracted function of the division), it is quite possible for the specialist to pursue the work adequately without ever coming near to an understanding of the software's application. In the other 'new' area, controls performance, which deals with stability and overall design quality, a generalist expertise and a 'systems' view is essential. This must take in microelectronics since the latter now forms around 25 per cent of the final product.

Graduate specialists in the division thus numbered 105. In addition there were eight technicians and eleven administrative staff, making a total of 124. Out of the turnover of around £32m, over £30m went on equipment and £1.3m on salaries. From this we draw a picture of a high-capitalised but low-paid industry. It was noted, through discussion with secretaries, that administrative staff were comparatively well-paid – several had remained there for five years and more, saying they could not find better rates outside. This may be contrasted with the graduate engineers whose turnover had until recently been between 15 per cent and 20 per cent, significantly higher than the company-wide norm of 10 per cent to 15 per cent (which had been considered a problem).

Around half the number of technicians were employed compared with fifteen years ago. The director hoped in future to employ a hybrid of technical assistant/clerical worker which, he said, would be more effective than the current division of labour.

Skills and training

An emphasis was placed on the importance of long-term experience as a qualification, saying that in this industry you had done your best work by the age of 35 such was the pace of technological advance. This contrasted with the amount of time it took to integrate graduate engineers or managers into the company at the outset. Because the company business was so specific compared to the training received in universities, the rule was that it would be six months to a year before a new entrant was listened to and two years before anything they said was likely to be acted upon. We can thus see how there was a mismatch between the general skills taught in universities and the specific ones required by the company. This was not, of course, seen as a shortcoming of the education system, since it could not be expected to provide L5 with exactly what it wanted. The Technology Support director said he would much prefer to recruit people largely via sandwich courses so that they would be of immediate use to the company once their education had finished. The paradox of the whole process, as far as Technology Support was concerned, was the 'specific' of L5, as opposed to the 'general' of the order in itself – the graduate had to understand how his specific discipline fitted into the complicated whole.

The director found that the people he employed often lacked the expertise to monitor whole systems. He himself had been trained as a systems engineer but did not see many people who wanted to go in that direction. Through the sandwich-course route he hoped to remedy this, although he had not yet won the argument – he referred to 'others' who would argue that what they wanted was more specialised people. Again we may point up the problem of diversity and integration. Another manager had referred to the impossibility of any person ever being able to arrive at an overview of all the disciplines now involved in system building – the delays to communication alone were almost insurmountable. Not only had the number of disciplines expanded, but their inputs had become more complex.

It was asked whether it would be possible to get round skill shortages by training technicians to carry out graduate jobs. Apparently this would only be possible in around 5 per cent of the tasks to be covered, which represented only a small grey area filled usually by lower-quality specialists. Rather than providing a service to engineers, technicians in this division carry out these 'lower quality' tasks themselves, such as operating the environmental testing facility, which involves work of a more routine nature. A general technical/clerical assistant would, however, smooth the working of the various groups of specialists.

With regard to recruitment and turnover, Technology Support was now in a bizarre position – namely that turnover was not high enough. The CAD/CAM project manager had explained how company turnover, which stood at '10 per cent to 15 per cent, was a problem – 5 per cent was seen as a natural level. In the more specialist division, turnover normally stood at around 15 per cent to 20 per cent which reflected something gravely wrong either in selection or word/reward organisation. Recently, however, it had shrunk to between 5 per cent and 10 per cent. In a specialist unit this was unacceptably low – the director was worried lest it should transpire that all the good people had gone elsewhere. The phenomenon was too recent for a conclusion to be drawn. Turnover was closely related to the basis for recruitment and training. L5 itself, with its emphasis on the company-specific, did not, for its purposes, require that graduates coming into the company pursue training in accordance with the attainment of membership of an engineering institution. The latter takes around eighteen months to accomplish and the opportunity for doing so is often a necessary enticement to get people into the company – who according to the director could get by with a less extended and elaborate formal training syllabus. A figure of three months, involving movement round different aspects of the process, was seen as sufficient. The problem was that, while institution-guided training was often necessary to attract people, it also made it easier for them to leave the company after its completion – around 25 five per cent to 30 per cent of the institution trainees left the company immediately on completing the course, after which the rate of departure dropped to a steady 5 per cent to 7 per cent.

Institution membership seemed to be a catalyst for hydro-mechanical engineers leaving rather than electronics – in the case of the latter, the cause of departure was more often frustration at not being able to design to anything like the extent they would have liked, and for which their undergraduate training had prepared them. Instead, they were largely there to check and monitor the design work carried out by suppliers. To meet the expectations of electronics engineers, some kind of electronics manufacturing plant would have to be established on the site.

One of the difficulties which would deter the company from making such a move, would be in the area of recruiting enough people with electronics skill, at all levels, into a company where there was no tradition of work in their area. Skill shortage thus threatened the capacity of the company to increase its electronics effectiveness from two distinct angles – without going into the business wholesale they would have trouble finding people, while the national shortage of such people would augur against any such wholesale move.

Software assurance

The most critical area, which had caused problems, is that of software assurance, which had become increasingly important. Only 10 per cent is actually done within the company while the rest is put out. The shortage of qualified people restricts the amount of control that can be effectively exercised, although the 'independence' of software makes it the most easily sub-contracted of the specialist checking activities, no engine awareness being necessary. To lighten the task software was divided into three categories – critical, essential, and non-essential – with critical receiving the most attention, since 'bugs' in this area will lead automatically to fatal accidents. The independence of software is counter-balanced by its invisibility – mistakes are far more difficult to trace (the case of Company M4 where considerable trouble was caused when, after a certain level of product complexity had been attained, it became almost impossible to tell whether a fault lay in hardware or software, functions covered by two distinct groups within engineering) – the safety requirements of aerospace necessitating the removal of all faults before this stage was reached. Attempts had been made to attract software engineers by paying an increment of £300–400 over and above the then basic engineer's starting salary of £9,000 – this calculation was based on seeing where people who left the company went to. This probably went hand in hand with the reorganisation of software engineers into a self-contained group. The process of assurance is highly complex and based on the document DOI78 containing recommendations for software. The documentation for a project's software could make a pile a metre high according to the director. These factors made software assurance the biggest disadvantage both to the sub-contract systems and the use of microelectronics in general.

The company was beginning to investigate the available commercial software more thoroughly, thus also cutting down on personnel requirements and sub-contracting costs. In another case we were told that the going rate for software contracting was then £30 an hour at the time, possibly over a period of several months. In view of this it is not surprising that the division took the view that, 'there's no point in us re-inventing the wheel'.

L6: 'COMPUTERS LTD'

General

Case L6 was a major UK site of a large American multi-national, involved in the computer industry producing very large batches. The site is divided into two distinct parts. The larger part of the site (L6.1) is involved with sales, marketing and administration for the company in Europe. The smaller part (L6.2) works quite separately and is largely devoted to design and development. In this, it is linked on the one hand to manufacturing plants elsewhere in the British Isles, and on the other to the company's US headquarters for its overall direction. Headquarters gives the final say regarding all projects undertaken at L6.2 and is also in a position to direct skills and training policy in line with the projects undertaken. The plant's managers none the less see themselves as more or less autonomous.

Personnel

The site as a whole had 1,200 employees, of whom 400 worked in L6.2. L6.2 was, in turn, divided between four divisions, all of which function as separate profit-centres. These divisions were:

1 *Networks and Communications (NAC)*
NAC is the fastest-growing and highest-qualified of the divisions. Its sole products are printed circuit board designs for inter-computer interfaces. The company produces these in tens of thousands enabling networks of computers to be constructed. The computers that are thus able to 'talk to each other' include those of major competitors as well as the company's own. NAC directly employed 110 personnel, all of them graduates.
2 *Information Office Systems Group (IOSG)*
IOSG is concerned with all-in software packages. Like NAC it employs almost wholly graduates (well over 100), although generally of a slightly lower standard. Again, like NAC it is geared to large-scale 'off-the-shelf' solutions.
3 *International Products Group (IPG)*
IPG is a very small division dealing almost entirely with the translation of American software into foreign languages, notably German and Russian.
4 *Customer Services Support Engineering (CSSE)*
CSSE employed around fifty people, including a large proportion of graduates (less, however, than the other divisions). It liaises with the user-base, providing a twelve-month post-installation service. It is linked to the other divisions through their technical writers. It does not undertake actual repair work, which is carried out separately from L6.1.

Also sharing the site on the L6.2 side was a further group, *Computer Special Systems (CSS)* which, since it represented a sub-site in itself, involved in small-scale and customised solution, is discussed separately (see pp.95–6).

Skills and training

For this review, we have concentrated solely on NAC since it best contributes to our understanding of the trends and counter-trends occurring in the electronics industry. NAC, like the computer industry in general, goes against much of what has been said, in this report and elsewhere, concerning a shift to smaller batches and customisation. NAC is geared to homogeneous batches going into the tens of thousands, although there is a paradox to the extent that these standardised solutions may mean it is easier to construct systems geared to the specific needs of the customers.

The insistence on large batches reflects economic circumstances and has important implications for how the business has to be worked. The company as a whole does not see itself as being in the market for customisation and feels considerable pressure to achieve economies of scale on the traditional model. Furthermore, according to the company, its customers nearly always demand off-the-shelf packages. This is particularly important where large systems are being constructed on the premises of major clients, using computers brought both from this company and its major competitors. The company's structure world-wide reflects this strategy. Manufacturing plants are situated hundreds or even thousands of miles away from where products are designed. This gives flexibility regarding the size, rather than the nature of output, excess capacity in one production plant can draw on surplus orders at another. One result of the strategy as far as design in NAC is concerned is that the process is split into two phases. The major development is done by NAC, whilst last-stage design and modification is carried out at the manufacturing plants. The use of this capability has, very recently, led to the inevitable result that manufacturing plants now carry out some of their own projects right from the design stage. Part of this overall strategy of streamlining with a view to cost reduction has been the hiving off of any potential for small-batch and customised development onto CSS, established as a sub-company internationally for this purpose.

What we are left with, as far as NAC is concerned, is an elite design and development division drawing on highly specialised staff, divorced from the 'give-and-take' involved in last-stage design, customisation, modification, installation or production. This arrangement, potentially at least, raises problems of its own where recruitment, development and training are concerned.

NAC's employees can be broken down as follows:

Hardware design	40
Software design	30
Project development	7
Senior project leaders	3
Computer-aided engineering	12
Lab technicians	2

The remainder is made up of management and administration, plus small functions such as quality assurance and technical writing. As is clear from the above, design is largely carried out by project teams, each printed circuit board being assigned a particular design team, including both hardware and software. Seven of the project designers have broader skills and work on a number of projects.

Not only is NAC entirely staffed by graduates, but these are more highly-qualified graduates than those in the other divisions (or nearly all of the other firms in the sample). Advertisements for NAC posts always reflected the company's preference for first-class honours graduates. Usually these were the ones recruited. One third of NAC's employees had first-class degrees, including most of the division's new graduates. Those staff that did not usually had the right kind of experience.

It will be noticed that technician-level employment was virtually extinct in the division, and on the site in general. This was a direct result of technological change coupled with the emphasis on larger batches. Technicians had in the past been employed for such tasks as PCB fault diagnosis. Over the last period, economies of scale had apparently rendered such a function meaningless. This was compounded by a hardware to software shift regarding the company's source of profits, a shift that had occurred over the same period. The use of chips had contributed to this trend by simplifying and reducing the number of hardware components. The company did not design its own chips, but did carry out quality assurance on those it bought in. Layer boards are bought in direct from the company's US plants. The buying of custom-built boards was likely to reduce costs further. These trends are expected to become more pronounced with the arrival of the application-specific chip with a mere ten components. This change will accelerate the convergence of hardware and software design expertise that was already seen to be occurring.

When hardware had accounted for most of the company's profit, hardware and software had been quite distinct. Now that the position had been very rapidly reversed the company was looking not merely for expertise spanning the two categories (which could no longer effectively 'avoid' one another), it was also seeking a more 'systems-oriented approach'. This point is where it encountered a measure of paradox in its skills strategy. Given that it regarded its personnel as its most important resource and that it had very strong technical capabilities and traditions, which it regarded as crucial to its competitiveness, it tried, as we say above, to recruit the most highly qualified recruits it could find. This, given the nature of the education system, meant it was recruiting the people who had proved most successful within a narrowly defined speciality. In turn, it clashed with the need to find people who could work in both software and hardware. Conversion courses were not seen as viable at this level; the company had to have people who had been trained to the highest standard in the disciplines they were going to apply. The company had apparently become more effective by straddling both hardware and software, and the position was seen to have improved in recent years. One of the means by which the division attempted to gain greater contact with universities was through joint research ventures; it was then involved with six major educational institutions on this basis.

Such ventures may have contributed to one of the abiding concerns of the company – how to keep its employees content. Having recruited many of the best people in the subject area, the company was aware of the need to provide them with consistently interesting and challenging work. This went beyond the practice of the company as a whole, which was to offer very attractive pay and conditions to its staff in order to keep them within the fold. In return for a higher standard of living, the company did, however, expect employees to show a correspondingly high level of commitment. In terms of peak work-load, extra labour was not contracted in, instead those working on the relevant tasks were expected to be prepared to work till 10 p.m. if necessary, although at other times they would rarely work after 5 p.m. These combined approaches appeared to be successful. The graduate rate of turnover was only 4 per cent annually. However, the personnel department had begun to break with tradition by attempting to establish a four year sandwich programme which, as well as helping to socialise recruits into the company, was expected to secure their loyalty in the longer term.

The question of socialisation into the company is related, as in several other companies studied, to the 'generalist/specialist' question. The company wants people who will be useful for more than just one job. This underlies the logic of project organisation. A number of people, whilst brilliant in their own specialism, were seen as unable to effectively see beyond it, whilst others, sometimes less outstanding proved to be more flexible. This distinction formed the rationale of the division having two parallel hierachies (of M6), one technical, one managerial. The technical hierarchy existed to provide consultancy services to product groups and to managers. It covered about one in ten of NAC's employees, including some who were said by one manager to be the type of people 'you can lock away and occasionally throw a banana to keep them happy'. Meanwhile the division was expending resources on improving the quality of the management hierarchy. Several engineers were sent each year for sponsored M.Sc. degrees at major business school. In-house management development was provided on the site. Short technical courses were also provided when development projects were seen to make them necessary. Of NAC employees, 10 per cent attended them each year. The company provided its employees with an active tuition refund policy.

Skill 'portfolios' and their development were approached strategically. At least once a year, a senior woman manager would visit the site from the US headquarters, and would carry out a skill needs analysis from which training course requirements would be deduced. A skills-register of employees was regularly updated. Both technical and skill developments were seen as a matter of establishing a base of expertise, consolidating and adding to it with each new project. Two entire projects were launched each year. Recruitment and training needs were geared specifically to what projects were being, or were about to be, undertaken. Skill shortages would never lead to the abandonment of a project. The availability of the right skills was a central consideration in the adoption of a project – 'you don't go for a project that you may not be able to complete; that way you would lose credibility'.

Each year NAC takes in four new graduates, generally two from electronics hardware and two from software. These would receive three to six months' company induction-training, learning the various operating systems. During the quarter when our interviews took place, twelve people were being sought. These included four new graduates and eight graduates with at least two years' experience. In addition they were seeking a CAE manager (they had recently taken a CAD manager from computer special systems on the same overall site) and a technical writer. Unusually, they were encountering difficulties in filling these vacancies, and were embarking as a result on a more extensive campaign using improved advertising in a larger section of the national and technical press. The four-year sandwich programme was being advanced as a way of easing these problems in supply.

Computer Special Systems (CSS)

CSS was situated on the second L6 site where it constituted an organisation in itself. It was part of a world-wide sub-organisation, each site of it being termed a business group. All business aspects were accounted for on the site or in its two branches in Europe. Overall, it was accountable to the CSS group within L6's parent company. The site employed 120 people (plus forty on the overseas sites). The object was to provide customer-specific rather than the off-the-shelf solutions provided by NAC and IOSG etc., within L6. For this customers had to pay an engineering premium on top of the software generated profit. The business group was twenty years old, but like the rest of L6 had only been on this site for four years. Under market pressure, the true customising aspects of CSS had declined. Only 50 per cent of its people or products would be termed customised. The rest were base products which had found market niches which required more investment but which were more profitable. By the mid-1980s, they had grown from 100 people in this group as a whole, to 160. Turnover had doubled from £50m to £100m. With the recession of 1981, they had drawn back from one-off customers and had moved to 'repeat customisation'. At one point, they were making 60–70 per cent base products. With recovery, the pendulum had begun to swing back again. Even customised products are made in large batches – a thousand units per large customer (such as a bank) for example. Although manufacturing is carried out on other sites of the L6 group, CSS builds it own prototypes. Of 160 only eighty people are in design/development. There are twelve in administration plus marketing product management, custom product management. This leaves forty people in actual manufacturing. Among engineers (who are organised into projects under ten project managers) there is a 5 per cent turnover. Around fifteen out of the eighty engineers are contracted in. In addition there are two students being sponsored at a local university. Of the eighty, only four are not graduates. Almost all are electronics or software plus some physics and maths and only one or two mechanical. In the past, they had an aversion to new graduates. Now, they sponsor in order to guarantee company suitability. It is difficult for them to compete for new graduates since they cannot offer as much as the larger organisations on the site.

The system of matrix management practised is so complex that an entire CSS training course with ten modules has been devised to explain it to new entrants. The course takes the place of one week's induction. The CSS system is used by the company world-wide. Part of the complexity besetting work in the design/development area (where the engineers are divided into stable groups of twenty in addition to projects) results from the fact that they did away with the 'soft'/'hard' electronics distinction four years ago. The integration of the two functions in customising equipment is itself particularly complex. The company has recently introduced CAE facilities of its own, although linked with those of L6, who recently stole their CAD manager.

A major effort is being expended to instil 'financial awareness' into the engineers, since in customising they are somewhat freer in how they approach problem solving (customers are not given quotes) while the profit margin may be more fragile. CSS want all its staff to understand how their contribution fits into the work of the group as a whole. Unlike the larger concern, CSS tends to have less division of labour and de-coupling of functions.

Unlike the larger company, CSS did not like sending people away on long courses outside of the L6 group itself, the training facilities of which were used. The manpower budget seemed tighter overall, although, as in L6 generally, salaries were high.

L7: 'MEASURING LTD'

General

Case L7 was a major site of an American multi-national. It was concerned with the design and manufacture of computerised measuring equipment and related products. An outgrowth had been founded on an adjoining site to constitute a microwave division.

The company was growing in employment. By the mid-1980s it had grown from 800 to 1,100. Over 250 of these worked in the new microwave division.

Personnel

The 800 employees in the main division consisted of the following:

Administration	70
Research and Development	110
Marketing	80
Quality Assurance	40
Manufacturing	500

The annual turnover was $100 million dollars. The company worldwide has been growing at the rate of 20 per cent per annum and 50 per cent of employees were professional or administrative. The trend has been for employment to move from manufacturing to the design process, with manufacturing itself becoming professionalised.

All R&D staff are graduates, 70 per cent of quality assurance, 60 per cent of marketing, 12 per cent of manufacturing. The overall figure for graduates is thus around 30 per cent. This has coincided with product technology developments whereby there are less parts, greater inherent reliability, microprocessor control and less assembly, although it was claimed that more dexterity was required for the assembly. Current products required only 47 per cent of the manufacturing cost of earlier models. Some of the final assembly and test work was put out to sub-contractors. The number of components had fallen by 60 per cent, from 1,000 to 400. Much infrastructural work, such as sheet-metal painting and cables, was contracted out so that the company could concentrate on what they did best. Manufacturing was in small batches – fifty units or less, and was design-intensive.

Skills and training

As far as graduate recruitment was concerned, the company claimed no problems whatsoever, due to their image and location (compare this with 'Rangefinders' – see p.173). Sixteen graduates had been taken on, eleven for the main factor and five for the microwave division. There has been an increase in the number of software graduates recruited, which has coincided with an increase in women graduates, six out of sixteen of the present intake. Recruitment is carried out cautiously – the company works to a five-year plan and has a no-redundancy policy – when work temporarily builds up, recruitment will not rise proportionately. The last year or two have been spent in a harsh business climate – normally twenty-five graduates would have been taken on. There is no precise plan for recruitment but it follows in the shadow of the long-term plans, and the fact that people must be brought in. At the same time, there is a problem arising from the rapid advance in technology. The 'half-life' of an electronics graduate recruit is said to be five years, that for a computer science graduate only three years. A retraining cycle is thus set in motion. In the mid-1980s, nearly £1m was spent in adult training on the site (including the cost of time spent on courses). The training involved 800 employees. In addition, people are being sponsored on college courses on different bases, up to 100 at any one time, with Ph.Ds included. Supervisory training played a large part. Since 1980, quality circles had been introduced throughout the company with the same emphasis on satisfying internal 'customers' and 'total process involvement' as characterises this and similar companies in the US. There was a similar emphasis on 'right first time' which at the technical level is allied to an expansion of CAD.

The company has, by putting money up front, achieved discounts for specially designed courses at local colleges and universities. Supervisors are expected to make development plans for their subordinates, taking all these options into account. Part of the aim of the higher training facilities is to keep employee turnover low – it stands at only 5 per cent to 6 per cent (including transfers within the larger company).

Once, all R&D was hardware-based. Later, however, it was 50–50 hard-soft. This has since moved to 60–40. The 'smart boxes' produced on the site for the telecommunications industry are easier to operate for customers, due to the higher level of in-built programming. The emphasis on built-in quality has led the company to educate its suppliers to reduce lead times.

Around 50 per cent of the company's products are concerned with computerised calculation, and 40 per cent with electronic measuring equipment and oscilloscopes. The predominance of PCB design in the company's value-added (although mechanical design is still important) has led to the company looking at expanding their CAD/CAM facility which they have been slow at developing in the past.

The microwave division is concerned with the development and manufacture of spectrum analysers. So far, the design is still being done in the US with customisation being down on this site. This is partly so that those concerned can be trained in how to design them. It has ten people in R&D only.

The site as a whole takes on twelve technical trainees a year. These are school-leavers whose hands-on experience at the company is combined with courses at the local college, leading up to HND level. After finishing the course they have one year's company induction. Over the three years of the course, there will thus be thirty-six trainees at any one time.

Chapter 6

Case studies: medium-sized sites

M1: 'MATERIALS LTD'

General

Case M1 was involved in the design and production of precision testing equipment. It was the major site (outside of the United States) of an American multi-national which has outlets in most parts of the world. The site provided direction for the company's activities throughout Europe, Africa and most of Asia. At the time, its manufacturing role had increased over the last two or three years when the worsening relations of sterling to the dollar made it more economical to manufacture components for new models on-site in Britain rather than import them from the parent company in North America.

The company has significantly improved its trading position through an increase in microprocessor involvement in its product range (a process which has been occurring over the past ten years), and further through a shift to software-driven machines. New process technology (notably CAD/CAM) is expected to lead to further improvements.

The company's products are divided into two divisions, the Structures Division and the Machines Division. Both possess separate design and production facilities on site, as well as separate sales and marketing. The first is concerned with large-scale systems, purpose-built from one-off designs for use in heavy engineering contexts. This range was seen as incurring unacceptable losses and has been cut back in importance in recent years. The Machines Division, on the other hand, has been growing in importance with several new ranges being launched over the last two or three years (recouping the ground lost by older models and the Structures Division). Machines Division products are designed for general purpose testing.

Whilst the Machines Division products fall into three standard categories, they are small and modular in design and can thus be customised to a wide range of applications. They are now designed to be operated by computer, the computers being brought in from outside, although the software is produced by M1.

Whilst the products of the Machines Division are produced on a continuous basis, with new varieties being launched at least once a year, the Structures Division is geared to specific orders and small batch or unit production entailing some uncertainty.

It is significant that both divisions have their own design, production and marketing facilities on the same site (in contrast with L1 where three product groups shared the same manufacturing facilities). Case M1 appeared to place considerable importance on close collaboration between the three functions, on account of the highly competitive markets in which their products were situated. There was an increased emphasis on design capability, within practical limits – the designers for the Machines Division approximating to a general product-design department, whilst those in the Structures Division were oriented towards meeting specific customer demands. The manufacturing facilities covered a wide range of techniques. These include automatic machine tools for precision components, computerised inventory and production control, and software-driven machinery for the insertion of integrated circuits into printed circuit boards. The testing department has become sophisticated as a result of these advances in technology. Computerised checking is in use, CAD/CAM is being implemented currently. There is also computerised inventory control, CNC machine tools for precision components, computerised production control, and software-driven machinery for the insertion of integrated circuits into printed circuit boards. Testing has moved to a higher level of technological sophistication.

Personnel

The company had 412 employees on the site. They included:

Engineers (including Design and R&D)	80
Production and planning staff	30–40
Operators	110

The remaining employees were involved in Sales, Administration, Services, etc. The engineers cover the whole range of mechanical, electrical, electronics and software, and draughtsmen. Of the 80, about 9 were software engineers and fifteen electronics engineers. The remainder also include senior scientists. Engineers are nearly always graduates taken direct from universities originally, often via the 'milk-round'. Engineers provide most of the complement of senior management and the sales and marketing functions (these categories are not included in the figure of eighty working engineers).

Skills and training

Different engineering skills have been in short supply at different times. Ten years before there was a shortage of mechanical draughtsmen. Later, there was a shortage of electronic engineers. The company appeared to have been able to acquire all the engineers it required, however, the prestige of the company and the efforts it put into marketing its image to graduates being seen as both necessary and decisive in this.

With more traditional engineering disciplines, there appeared to have been little problem regarding excessive turnover of personnel. Engineers were seen as contented to stay where they were in many cases. Software engineers, on the other hand, were considered to be less involved in the company and its products and more in software for its own sake. They were likely to wish to move in order to work in other languages, for instance. They might also switch companies for higher pay, usually after two or three years (this applied to other engineers as well). The personnel manager did not regret this particularly, or consider that higher pay would necessarily improve matters, since this could lead to stagnation. The company seemed to prefer to develop its own personnel from scratch and promote from within wherever possible. Attempts seemed to be made to inculcate a strong sense of company identity among the staff – a wide variety of social activities were organised through the personnel department.

As already said, most engineers arriving at the company were recent graduates in their particular field. The company did not believe in conversion courses to turn physicists, for example, into electronics engineers.

Conversion courses were, however, provided at the site for existing engineers who had been trained in other disciplines in order for them to be given the awareness of microelectronics. This applies particularly to technicians (not included in numbers given above), notably those working in the product testing department. Broadening of engineering skills was found to be practical in most cases, except perhaps when the engineers or technicians in question were over 50 years old.

The testing department was the area in which the company saw experience as being more important than qualifications. Although the test engineer himself would be a graduate, technician-level engineers played a major role here. A technician with years of experience was seen as more likely to spot when a fault was caused not by some integral part of the system being tested but by some more elementary factor such as the cable being in the wrong position and causing interference.

The Test Department was the route frequently taken by technicians in search of promotion to management or sales or marketing. Once a technician had a full certificate in technical engineering he was qualified to enter the Test Department. The company policy was to keep experienced test personnel in the company, which is perhaps a reflection on the increased importance of reliability in sophisticated equipment. The relation of the Test Department to After Sales Service has changed with advances in technology. Previously, almost all repairs and tests on customers' equipment bought from the company had been carried out on customers' premises. Now, a larger proportion of major testing and repairs is carried out on-site. This means that the service workers out in the field may now be less skilled than before. The technology means that it may be a question of some elementary diagnosis and the removal of a black box, which is sent back to headquarters. Over the last few years, field service has become less the preserve of engineers and more that of technicians. There are now more personnel employed in testing at the main site, but not more specialists as such.

Engineers are elected for their ability to work in a team. Engineering work is divided up into teams of six to eight engineers. Product development initiative usually comes through the Marketing Department, which finds out what the customers' needs are in advance. This appears to have been the case with the new products referred to above. Although many components are bought in, all development work is carried out by the company itself.

The main changes over the previous ten years have been the arrival of microelectronics and (over the past half-decade), the move towards software-driven machines.

Overall, a wide variety of training and related activities are sponsored or encouraged by the company. Technicians and others are encouraged to take part in vocational training projects at the local college. Engineers and technicians have been encouraged to attend electronics awareness courses one night a week. Apprenticeships and on-the-bench training are still widely used. Middle management are sent on management courses at Ashridge about once a year (not always the same people). Every time a new product is launched (around once a year or more often), the sales and marketing personnel are trained on the site in the selling of that product. The personnel manager has taken a governorship of the local college and has been campaigning to get co-operation from local schools to present engineering careers in a more flattering light.

The company also carries out research work in co-operation with universities. Employees were said to be given full encouragement to participate in Open University courses in a wide range of subjects.

The M1 Training Centre, founded back in 1972, provides training courses in the company's products for its own employees, including sales and service staff, as well as training the personnel of user companies to minimise machine downtime, via its Operating and Maintenance course.

M2: 'OFFICE LTD'

General

In case M2, the company was involved in the design, manufacture and marketing of networking, communications and office automation systems. It had started out as a mini-computer manufacturer twenty years before, but was now largely concerned with systems design (largely software and highly customised). Manufacturing was largely confined to final assembly and testing with most components and sub-assemblies bought in modular form.

As a British off-shoot of an American multi-national, the site visited was the only one concerned with design and production, the remaining two sites being much smaller and involved with business administration and sales.

Advances in technology had pushed the company away from traditional manufacturing and into design-intensive areas. It had also an emphasis on large projects, tailored to customers' requirements, although more standard product ranges were still important.

One project for a communications system with the City of London, providing a database of financial services information, was responsible for 31 per cent of turnover in 1985 (as opposed to 34 per cent in 1984). Over 35 per cent of turnover was provided by three other systems, with services providing the remaining 24 per cent of turnover. The City of London system has been sold to other large customers and was expected to provide a steady source of revenue in the future. All the systems had been available, subject to customising and modification, for more than two years. Systems for interworking already existing systems were the most recent major development and were providing an increasing percentage of turnover, although the precise amount is not clear from the accounts due to their being grouped with one older system with which they are habitually used.

Thus, as with L1, one reliable system-range will continue to draw steady revenue, with increasingly sophisticated network-interworking systems (involving more significant development and customising lead-time) gradually taking on a more important role.

Production is far from being automated. The company does not even use its own products in its offices. In the production process, the only major high technology resides in visual-aided assembly for printed circuits (where the position for each component to be inserted is highlighted in advance as with painting by numbers) and in the computerised testing facility which was installed two years previously. The testing facility, at a cost of £250,000 would not normally be considered an economic proposition for such a company. It was said to have been installed in order to ensure the highest standard of quality in the company's systems, the latter being sold on grounds of quality and adaptability rather than low price, which was not seen as the major priority (see our comments under 'Marketing', however).

Personnel

The composition was as follows:

Engineering	
Engineers (HNC)	110
Development engineers	120
Procurement (mostly HNC)	50
Total	280
	(half of labour force)

Half of the 280 were engineers, half technicians.

Overall	
Sales and support	120
Engineering (inc. manufacturing)	330
Marketing	25
Commercial	50
Personnel	5
Remainder in administration	(120)
Total	650

The total had not changed radically in composition or size, except for slight increases in the area of development.

Skills and training

As with many companies covered in the study, M2's strategy meant an emphasis on graduate recruitment and development.

As far as skill shortages are concerned, the company claimed it had noticed them but had managed to avoid being seriously affected. Graduate recruitment (fourteen a year, unless specific expansion meant a larger intake), was the only area which could cause problems. Graduate turnover, at 20 per cent, was high but appeared to be falling. The company appeared to have devoted considerable energy to induction of new recruits, with all senior management taking part.

In general the policy is for training to take place within the company. The in-house training facility has a budget of £300,000. The company is not as good at training for the shop-floor level as it is for those higher up. Supervisors are sent on customised courses at a local polytechnic. Engineering apprentices are sent to local colleges on day-release. MBAs are sponsored through business schools, where senior managers also attend courses. Otherwise on-the-job training was the norm, although some conversion training had been given to field-service engineers. The skills repercussions of the advances in product technology were complex and can best be summarised by reference to each function in turn.

Research and Development

The main effect here has been a considerable expansion of the department, with a switch from hardware to software in its emphasis. Most graduates (80 per cent) had graduated either in electronics hardware or software (the precise proportions are not available yet). Most graduates are recruited into the development department, where a majority of those working are graduates.

Poor marketing (see pp. 106–7) threatened the competitive edge provided by the company's strong R&D capability.

Production

Production has tended to become simpler with an increasing proportion of system hardware being bought in an advanced state of assembly. As with L1, a number of workers (mostly women) are retained in the PCB assembly process, although as in Office Ltd these had been deskilled by the 'painting by numbers' equipment. Printed circuit boxes assembly itself has become somewhat marginalised as the company moves from buying components to buying assemblies. As stated above, the value-added relies chiefly in the software. This development has apparently been the result of increasing reliability of microelectronic systems generally, in the sense that distinguishing qualities have to be of an increasingly 'higher' order.

Quality control

In contrast to production, quality control has, if anything, expanded in its importance – and to some extent in the skills involved – with the introduction of the expensive computerised testing equipment described above. This would represent a skill shift away from 'craft' testing, towards straightforward computer ability. It might be necessary to enquire as to the level of skill now required and how it compares to that in use before. As a function at least, quality control appeared to have taken on a greater importance, a result of the increasing reliability of micro- electronic systems, and the need to achieve accordingly higher standards in order to maintain a 'quality' image.

Marketing

Marketing appeared to be the 'weak link' in the company structure. According to the personnel manager, there had been a tendency in the past to have 'unreconstructed' engineers as sales and marketing personnel. There was an attempt to progress to produce a more 'professional' marketing department, with a smaller proportion of engineers. The difficulty may be traced with changed customer profile of the company, which, until the early to mid-1970s, had traded largely with universities and research institutions, the main product being the minicomputer. With the advance of microelectronics and increased competition, the company had been forced into a more commercial market place. The highly personal tradition of the company (which is non-unionised and 'paternalistic') had involved the marketing function's dependence on the managing director and his personal contacts. In theory, marketing had been the leading agent in new product generation. In practice, the 'research' tradition survived in the development department, where advances tended to happen unaided, and not according to a fixed plan. The development function thus appeared to be filling, in its own unstructured way, the vacuum left by marketing 'not doing their job'. This in turn led to a perceived need to control the enthusiasm and perfectionism of development engineers (whose graduate training was not seen as having prepared them adequately for work in market-oriented companies as opposed to pure or applied research). Senior management concerned over meeting customer deadlines, would intervene to ensure that modifications being carried out by development engineers were cost-effective. A modification that was unnecessary but which would enhance quality would be permitted if it took an extra day's work, but would be rejected if it were to require six months, for example. Development engineers were none the less equipped to monitor and evaluate what competitors were doing, even if this was seen as academic interest on their part.

Customer service

In some respects, marketing and customer service had merged in function. As with the other companies studied, products had to be adapted and customised to an increasingly large degree. Also the logical conclusion of the company's emphasis on systems software rather than hardware had been the expansion of software services and customer training to the extent that they accounted for just under a quarter of the company turnover in the mid-1980s. The company was thus becoming more skill-based and less technology-based.

As far as field services for hardware were concerned, a skill polarisation had taken place. Maintenance workers had been de-skilled to become unit-swappers or, at best, chip-swappers. On the other hand 'support engineers' in field services had been up-skilled to cover the area of software. This was brought about by converting existing engineers, through day-release etc., rather than recruiting software-trained personnel from outside (this would presumably have been too difficult or expensive for the status of the job concerned). The polarisation reflected the polarisation in the product itself, with hardware of greater external simplicity and software of greater complexity.

New staff had to be hired for development and marketing/sales. Support engineers had to learn new skills (as did sales personnel) who received an intensive three-day programme of training with the launch of each new product/range.

Quality control personnel had to adapt to new equipment, as did assembly personnel. In the case of the latter, redundancies had probably occurred, though no number was given.

M3: 'BATTLE COM. LTD'

General

Company M3 was involved with military communications. It had seen a gradual transition from mechanical to electronic engineering and a blue to white collar shift. A new managing director had set a large technical investment programme in motion since 1980.

Personnel

In 1981, the company had 380 employees. By 1985, this had risen to 512. It had not risen in proportion with growth. On the electrical side, for example, there had been major increases. The number now stood at sixty-two in direct production, and thirty-five in the electrical laboratories. Figures were not obtained for the more traditional mechanical side which was largely involved in the building and fitting of mobile containers. The composition of the workforce had altered in line with product orientation.

When the company started it had been 99 per cent blue collar. Now, white collar staff made up 40 per cent of the total. Turnover amongst staff stood at 6 per cent, while for blue collar it was 15–20 per cent. There were forty contract workers, mostly electrical craftsmen.

Difficulties occurred regarding the supply of electronics and software engineers, due to high demand in the labour market. Although the company had a policy of paying less than the market rate rather than rocking the boat on differentials (they were non-unionised), this policy was not applied to software engineers. Also, the site was not in a traditional industrial area – representations had to be made (often through customers) to depressed industrial areas such as Merseyside in order to acquire people with the right background and qualifications. Moving from such areas did entail problems such as sharp differences in the cost of living.

The company had so far been relatively successful in up-grading the skill-profile of the workforce. Like other companies they had begun to set up links with particular educational institutions. In this case a connection with a polytechnic came about by accident when an employee who was being sponsored was unable to get into a University. They found the course offered there – a sandwich course in communications – to be useful, and it was likely that it would be something of an apprenticeship route. A graduate would be required to serve three years in the company after being sponsored through. Two were following the course at the present time. Personnel plans were made each year and submitted by each department to the managing director who would then cut them down to fit in with overall budget. It was in the interest of each department to overstate their real needs and to insert unnecessary items to pad out the proposal so as to safeguard what was really necessary. Personnel would draw up the training plan after consulting with the engineering departments and would then cost it. The bargaining process could take anything from two weeks to three months. All personnel costs were recharged to the individual departments they served. They would have preferred to have had their own overall budget.

Skills and training

The change in the skill composition of the workforce reflects the changes that have taken place in the business orientation of the plant. Having started as a facility for refitting war-planes in the 1940s–50s, it began to specialise further in the direction of radio equipment servicing, a move which reflected the increased sophistication of in-flight communications. Later, the company carved out a niche in the mobile communications area. It had a secure market – 80 per cent of its sales went direct to the Ministry of Defence, the remainder (of which the proportion is increasing) going to other MOD suppliers. In recent years particularly, the traditional relationship between research and development and production seems to have been inverted to some extent. The development section could now survive on its own as a profit centre, selling its ideas (usually in specialised areas of electronics) to larger customers. Already an appreciable amount of production time is devoted to building prototypes. The staff changes that accompanied these moves were far-reaching.

First, the proportion of employees in the electronics as opposed to the mechanical side increased sharply. Second, the level of qualifications in electronics outstripped that of the mechanical division. Third, electronics acquired a growing development function which was largely absent in the mechanical division.

What occurred was an increasing gap, in terms of role, qualifications and strategic importance, between the two disciplines at the higher end of the scale. Paradoxically, the same sequence of events saw a merging of mechanical and electrical functions at the production and craft levels.

Up till 1980, only two or three electronic engineers had been employed in the laboratories; this has since risen to thirty-five. At the start, only one of the electronics personnel had a degree; now over half did. As mentioned above, in order to secure the supply of these graduates, the company sponsors students or employees through various institutions. About fifteen have gone through this route. Apart from Brighton Polytechnic there have been people sent on day release to Kingston Polytechnic or have finished courses at the Open University. The minimum acceptable for development work was an HNC or HND pass. In addition, there were two Ph.Ds and two M.Scs working in the labs. This reflected the increased importance of 'pure' development contracts (for the MOD) in the company's business. It was considered that such projects required a strong academic direction rather than more traditional styles of engineering management. Their personnel officer saw a change in the sense that authority in the labs was now based on respect for knowledge rather than on personality. No similar up-grading had occurred on the mechanical side. The kind of products for which such expertise is used or sold are in the areas of military radio, telex, radar, facsimile, and frequency-hopping.

Multi-skilling policies have been implemented lower down between mechanical and electronic departments not only on account of the complexity of the products to be maintained, but also because of the new production technologies that are being applied. Mechanical fitters, after the completion of their apprenticeship, undergo training in electronic maintenance – this in view of the need to maintain CNC machines on the shop-floor. Fitters are also trained in basic electronic service. The production engineer has also attended similar courses. On-the-job training is expected to account for more.

Much of the direct production work in electronics is concerned with PCB soldering and wiring, cable forming and fibre-optic links. This work is almost entirely the preserve of women (for reasons of dexterity and lower pay, as in other firms), of whom around thirty are employed in this capacity. Different batch sizes are produced, including prototypes for the development division. The work is semi-skilled and based largely on slowly acquired experience.

Women are hired at the age of 16 to 20 and take about nine months to learn the job, through practice with dummy boards, soldering irons, drawings and being attached to more experienced workers. Although no redundancies have resulted from miniaturisation etc. (the job loss is hidden in that increases in turnover have not meant any increase in the number of production workers), this area has seen some industrial relations difficulty. The work is multi-skilled in that operatives carry out most or all of the different functions – PCB assembly, cable-forming, etc. Discontent has arisen because of the greater simplicity, and hence boredom, of PCBs in recent years. Older workers particularly feel alienated in that they have been moved a step further away from the end product. In the 1940s and 1950s, one knew what was being produced and why. The personnel department had attempted to resolve the stream of complaints they had been receiving from the women through the introduction of a productivity scheme which they claimed had been successful.

Male workers in mechanical production received training to cope with CNC. This was generally of the on-the-job type and specific to the machine being used. There was a sense that the company was moving towards an increasingly sophisticated identity in the development and design area, which was expensive both in salaries and equipment. To get the pay-back from this investment, the remaining production facility had to be rendered more sophisticated and much more cost-effective at the same time. This process is accelerated by the need to get the pay-back on the investment made in the appropriate CNC, CAD/CAM equipment. As with L4, M3 appeared to wish to retain the maximum design and development capability in terms of graduates employed. To rationalise such work would be to lose the distinctive edge their products had in the market place. The result was a rapid movement in the direction of the design company employing 50 per cent graduates, with production retained partly in order to produce prototypes and small customised batches.

M4: 'MICRO LTD'

General

The company, in the case of M4, designed and produced electron microscopes, image analysers, compound crystal growth systems, E-Beam lithography, and a variety of products specifically developed for the semi-conductor industry. The electron microscope accounts for 50 per cent of a growing turnover. The company's prosperity is of recent origin. In 1980, it had been on the verge of total collapse. It was rescued by an entrepreneur who still owns and controls the company. Between 1983/4 and 1985/6, turnover increased from £25m to £50m.

Personnel

The company had 800 employees on site and another 400 world-wide. The latter were engaged in sales and services. In 1984 the site had 600 employees and in 1985, 700 employees. Since the rescue in 1980, administrative overheads have been kept to a minimum, as seen below.

General and administrative (not including directors and top management)

Finance	42
Secretarial	14
Personnel	7
Total	63

Personnel also included eight apprentices who are accredited to that department. Since they are daily assigned to production or engineering where they will eventually be permanent staff, they are not included here.

Engineering design and development

Engineers	94
Supervisors	6
Admin. support	7
Quality assurance engineers	37
Project managers	15
Crystal growth engineers	3
	162

Production		
	Machine shop	99
	Specials	24
	Assembly (mech. and elec.)	93
Test		83
	Supervision	42
	Support	78
		409

Others

Marketing	27
Sales	18
Service engineers (HND/HNC)	34 – technician HND/HNC
Management	2
Production support	23

Product training (for customers)	6
	110
Total	754
White collar	320
Blue collar	400

(These totals depend on whether service engineers are classed as white collar.)

The slight advantage of blue over white collar can be put down to the fact that despite its high-tech product profit, all the mechanical and metal cutting aspects of the product are carried out on site. Paradoxically, the fact that the high-tech products can only be made in very small batches means that it is not worth automating the mechanical semi-skilled side. Equally, it is more convenient to keep them on site rather than contract them out. In this unusual case, traditional blue-collar jobs are actually guaranteed by the success of extremely high-tech products. A substantial number of administrators and indirect personnel were got rid of in a slimming-down exercise. However, the total was lower because of a smaller volume of business.

Turnover was around 0 per cent in the production area. Forty-six new direct workers had been hired. In the engineering area, around forty were taken on. Turnover is generally low, although in the case of software engineers one is lost each month on average, making a turnover of around 10 per cent.

Skills and training

Although a large number of blue-collar workers were also needed and taken on, the only skill shortages occurred in the graduate engineering area. The engineering manager questioned said that the most difficult person to find was a good mechanical engineer. In this discipline, the company were usually looking not for a new graduate but for someone with experience. This was because their mechanical work tended to be particularly specialised in many cases. They were looking for people who had vacuum experience and who had been involved in precision work involving vibration of as little as one or two angstroms. These were hard to find, and, once found, they were prone to leaving for more money. It was noticeable that the average age of mechanical staff in design and development was as high as 48 while for electronics it was much lower (also in software for obvious reasons). While this did not cause the problems that might have been expected in implementing mechanical CAD (the suppliers were aghast at the thought), it does suggest that the company were able to keep people who were geared to spending their life in the company, but that they had problems in attracting and keeping newer, more specialised and career-oriented mechanical engineers.

Until recently,the company had inclined towards recruitment rather than training in order to meet skill requirements and, partly on account of their usually paying the market rate, they have been successful in significantly expanding their engineering department. Their paying of the market rate should not be stressed too much. The fact that the engineering manager said they were careful not to set off a never-ending spiral of differentials suggests that some constraints were operating in this area. Generally though the problem seemed to be a lack of suitable people in the job market, rather than any difficulty in attracting those that were there. An engineering manager said that, whereas in the past they had managed to fulfil their requirements through recruitment, they were now growing too big for that.

This change, of course, contrasts with what we have seen of the very large firms who were able to cream off large numbers of specialised staff from the market, whilst small companies were forced to train their own. This company is probably somewhere in the middle. While still small, but with high-tech products and a pleasant environment, they can compete with the largest companies. Once they pass a certain point, they may be faced with the worst of both worlds. They no longer possess the attractions of a small company, and are unable to provide the extended career. Furthermore, their size and the sophistication of their product make their skill requirements greater than may easily be met by anything less than a large-scale training programme.

The company presents the middle ground between the skills strategies of large and small firms. The image of the company and the high level of complexity involved in its products meant that it could compete with large companies on the 'milk-round' for new graduates. When it came to more experienced specialists it was less well placed, first because it could not offer an extended career structure and, more importantly, too few other companies were working in analogous areas who might have provided the appropriate experience. Company M4 was thus obliged to follow the 'small firm' route and develop its existing employees' skills on the job.

The accommodation of all-round skilled people from earlier years was a feature in the company. The small-batch mechanical production side of the company had been saved through the introduction of CNC equipment since 1980. It was accepted that the operators who pre-dated their introduction could not be expected to sit and watch them at work. For this reason, they had chosen machines which required greater discretion, and for which the operators chose their own programs. Compatibility with these machines had influenced the choice of CAD system.

In the engineering area, an elaborate form of project organisation had been established which seemed likely to have positive benefits regarding skill development. Overall supervision was minimal, consisting of two 'resource managers'. In this respect, M4 resembled L6, one of its customers, and similarly a firm with a long tradition as academic suppliers. Both emphasised self-supervision and individual commitment. The engineers (108) were divided into two groups, fifty working on projects, the rest constituting a 'pool'. There were 130 projects, lasting on average nine to eighteen months, with two or three developments being pursued in each at any one time. Some were concerned with new product-launches whilst the majority were involved with customised adaptations of existing product lines. Most new product development grew out of such customisations. Customisations occurred at an unprecedented level, with almost anything being considered feasible.

Thus, the project engineers were on average working on two to three projects at a time, whilst the eight project managers would be taking charge of an average of ten to twenty projects. Project managers are not chosen for technical brilliance but for their more generalist talents. The original notion was to develop specialist expertise on the projects (as had been the case with the product-based organisation that had previously existed in the company) and to use the 'pool' engineers as back-up. Thus pool-based engineers were generally more junior. More recently, the firm has begun to up-grade the pool in terms of skilled people assigned to it. In this way, a new stratum of 'generalised specialists' will be developed through using experienced people in a wide range of work.

Such measures are aimed at breaking the deadlock that threatens between different disciplines and specialities on the more elaborate products. The engineers are divided into mechanical, electronics hardware and computer science, with each accounting for around a third. Software is the fastest growing group and now accounts for the largest part of product's value-added. Software also made up the bulk of the forty graduate engineers recruited each year. A turnover problem existed in so far as new electronics hardware and software graduates regarded the company as good for a first placement for two or three years, but not large enough to offer suitable career opportunities. As products become more complex, the interfaces between the disciplines become less well defined and fault diagnosis becomes clouded, particularly where the respective responsibilities of hardware and software are concerned.

As the whole of the product becomes more complex, the customisation on which the firm's market share depended has to be based on increasingly precise information, otherwise its modification may be disastrously time-consuming. The introduction of CAD represents an attempt to generate more detailed and accurate information. At the same time, there was seen to be a need to move away from traditional 'trial and error' practices in engineering. For successive customisations for the same customer to be possible, the initial modification has to be scrupulously documented. Hence, a greater emphasis on being able to understand how the different parts of the product and the inputs of the different engineers all go together.

This emphasis on customisation and the integration of different disciplines occurs alongside a related development, a reduction in the ordering of customised components or sub-assemblies. It was the engineers' policy never to order anything that was not available from a catalogue. In this way it was hoped to standardise and simplify the parts of a product so that the engineers would be free to concentrate on the system as a whole, that, through elaborate software design, was becoming far more than the sum of its parts. This reflects the 'simplification/complexity' spiral of product microelectronics that we described in an earlier section (see pp. 56–9). To cope with these developments (which already meant that no one person could conceive of how all aspects of the product came together) the company was thinking of recruiting some experienced systems engineering expertise, in addition to developing generalist skills among its existing staff.

The major changes regarding skills were in the electronics area. Mechanical engineering seemed secure in its traditions, among which, interestingly, was the lack of a formal distinction between mechanical graduate engineers and mechanical technicians, which went alongside a less clear division of labour regarding design and draughting than that found in electronics.

In the electronics area meanwhile, a new shift was occurring. The company were about to move into the use of gate-arrays to increase product flexibility and a higher level of application of integrated circuits generally, some of which would be of the semi-customised type. This was seen as leading towards still more automated testing, and a lot less PCB layout work. It would also mean more electronics engineers (who were expensive) and fewer electronic draughtsmen (who were relatively cheap). These changes would not affect the mechanical side. This was and would remain fairly primitive, protected by the fact that low volumes could not justify modern techniques such as plastic moulding. Thus, it is in the area of the swiftest technical advance that the most dramatic changes in personnel were required, not in the older areas which one would have expected to become obsolete.

Similarly, older product areas, such as the electron microscope, whilst they became more complex, did not lead to the 'systems incomprehensibility' engendered by new product lines, since people had 'grown up' with the former and had had time to adjust to changes in its developments.

A more general transition was occurring regarding the culture of the firm. Now out of its crisis period its 'paternalist' and 'academic' traditions of informality were being slightly eroded by a new emphasis on written communication (regarding design cost/benefit for example) which had previously been minimal. To an extent, the company were 'suffering from success' and debating (so it seemed) whether their traditions of organisation were not to some extent incompatible with the growth in which they were engaged.

M5: 'TELCO LTD'

General

Case M5 was involved with designing and manufacturing teleprinters, data transmission equipment and anti-listening devices. The company performance has fluctuated. It had risen to £12m turnover, but then deteriorated. The distinguishing feature of the company's marketing has been its reliance on large-batch contracts to single large customers. Their teleprinters were designed and sold almost solely to British Telecom. When the latter was privatised, M5 was severely hit, since the new Telecom commercial management approach meant that large stocks of equipment would no longer be maintained, and orders would be lower. 'Telco's' response has been (a) to lean more heavily on their other large customer, the Ministry of Defence, and (b) to make moves in the direction of small batch operations via the factoring of other companies' products. Small batch work, as at L6, has led to a small outgrowth of the company being set up on the same site.

Personnel

Around 300 people were said to be employed by the company, although they appeared to have included in this figure a number of people who left due to the contraction of the business and technical advance.

Engineering	46
Administration	25
Indirects	10
Testing	20
Direct Labour	70
Sales, service	20
Factoring	40
Total	231

The last group was the branch of the company that sells, catalogues and factors other people's equipment. Half of its total can be counted as engineering and half as sales. They are engaged in the development of customised systems. The number of direct workers has been cut from over 100. In general, however, the loss of jobs either relatively or absolutely appeared less severe than in larger companies where, for example, turnover had doubled while the number of workers declined. Once, there were only thirty engineers. Later, however, there was already parity between mechanical and electronics engineers. The ratio is now 2 to 1.

Staff turnover is low – only a couple a year, of which one will always come from engineering. These were said to go for personal reasons rather than poaching. However, one of the reasons the company does not insist on graduate entrants is that they frequently lose their new graduates.

Skills and training

Engineering

The company were only taking on four or five people a year. There was no special graduate emphasis. Recruits range from ONC to HND to degree level. Pecking orders appeared to be informal, although ONC people would be more likely employed on CAD draughting (for which graduates are not used). Generally, one third of engineering recruits are graduates. These are acquired through the local press or through agencies. The reasons for advertising locally are that graduates in prosperous areas are unlikely to find the small scale of M5 appealing, whilst there are problems in importing people from more distant regions. One engineer was lost within weeks on account of the high house-price differentials between the areas concerned. The company have occasionally offered a package with bridging loans etc., to secure recruitment of such people. Because of the proximity of so many large companies requiring similar people, there have occasionally been no applicants at all.

Software expertise rather than software graduates are sought. The company likes to recruit 'all-rounders' – electronics engineers who can write software. This still leaves a gap where systems expertise is concerned. In all, there are only ten graduates in engineering, which suggests that all those who leave are graduates – that is, if two are recruited a year and one leaves then that would explain the low figure despite several years' recruiting. All ten of the graduates are employed in electronics development. Of these about a third also work in software development. The company are generally 'nervous' of software, partly because of skill shortages and partly perhaps because, for a small and traditional electronics company, a major move into software would constitute an unacceptable shift in organisational structures and culture. This nervousness held the company back from any large-scale linking of their various 'boxes' into systems which the huge software load entailed. Around twenty systems are produced by the company. All the software input is duplicated, presumably to minimise the pressure for an increase in software generating capacity. On the other hand, the same attitude may be seen in a positive light in the sense that it means that the company is, to this extent, tailoring its product strategy to suit its existing human resources (which they said was their greatest asset), rather than the other way round, which could be seen as wasteful. If they are to survive, companies of this type become dependent on subcontracting – M5's first reaction to new possibilities in the product is to say 'can we buy it'? The introduction of CAD has increased the possibilities here – it enables photoplotting and the digitising of holes for electronic components so that they are sent to the manufacturer in a more directly practical form.

Regarding graduates, we should add that one was sponsored each year. A problem arises when these succeed too well. One sponsored graduate attained a first in electronics and left M5 subsequently for a larger firm. 'Telco' accepted that they could not provide anyone with such qualifications with a satisfactory career structure. At any one time, the company will have three or four engineering trainees who start off in production.

Technicians and shop-floor

As with M4, there was a fairly indistinct line between engineers and technicians outside of electronics. Long-term service in the company was seen as more important than prior qualifications apart from key areas of electronics. Below this level, there were the more strictly technician grade jobs, notably testing. The company had around twenty testing technicians. The skills of these had become threatened by the introduction of automatic testing equipment and presumably a decline on the economic importance of diagnosing faults given the cheapness of microprocessor PCBs. Although the company was not unionised the technicians had managed to secure some retraining to broaden their skills, having grasped in advance the threat posed by ATE.

Since the introduction of microprocessors seven years before there had been wide-ranging changes regarding assembly work. First, specifications have become more complex and components and sub-assemblies smaller. These have gone beyond the limits of 'hand-layout'. This has brought CAD into the process strategy. CAD was introduced to cut lead-times and 'get to the market place six months earlier'. This need was allied to an increase in cost per component related to fewer bolted components. In turn, the adoption of surface-mounting techniques had facilitated the move to smaller and more complex arrangements. The results of this process are a shift of skill away from the shop-floor and into the drawing office and of production away from the plant and to sub-contractors. As we have already said, the introduction of CAD not only enables the company to design and draft complex miniaturised layouts more quickly (if indeed they could have been designed before), it also makes their assembly by sub-contractors cheaper. Thus, we can see how CAD facilitates the trend towards the geographical separation of design/draughting and production even as the appearance of CAD/CAM has provided the potential to bring them closer together.

CAD/sub-contracting has perhaps a dual effect on shop-floor skills. On the positive side it was seen to be leading to an emphasis on the 'packaging' aspects of the product which may require a renewed emphasis on mechanical skill and more systems awareness. On the negative side, many of the traditional skills associated with electronic assembly have virtually disappeared from the company. Hand wiring has gone and ribbon-cabling and cable-forming are much reduced. As we have seen elsewhere, the skills are not done away with altogether – there are still orders for old equipment which have to be met. With each technical advance labour content is seen to decrease. The result has been a gradual fall in the number of direct workers, who now make up barely 25 per cent of the workforce and who were said to use less of their skill than ever before. The company still paid the same wages as before to the de-skilled workers.

Jobs have been broken down with the adoption of surface mounting. The production services department prepares detailed instructions for assembly workers in a way which fragments the tasks involved. The company did little formal training. Its training budget, out of a turnover of £12m, was a mere £7,000. Beyond the basics of each job, employees were expected to educate themselves on the job. Apart from the test technicians, who pushed for training to have themselves up-graded, the only formal training appeared to be geared to engineers/managers. Every department would put in for training where possible. The company training officer was generally concerned with customer training.

M6: 'CONTROLS LTD'

General

Company M6 designed and manufactured control systems for a variety of industries. Their engineering department was divided into three product-based divisions.

1 Marine-control systems for vessels – engines, propellers, heat, communications. Customers: British Shipbuilders, Ministry of Defence.
2 Mining systems: M6 were one of the contractors employed by British Coal to develop and manufacture MINOS computer-based control systems.
3 Engine controls: This group was concerned with control equipment for aerospace engines.

One of the results of micro-processor-based technological advance has been the convergence of these groups and the principles involved in each. The triple hierarchy has largely been abandoned in engineering in favour of a project-oriented work organisation, a problem of the former arrangement being the waste and duplication engendered by engineers in each group not sharing the knowledge which was becoming increasingly applicable to all three. Flexibility has increased with technical advance-equipment which has been designed so that it may be used for control *and* monitoring or control *or* monitoring. It has also found applications away from its traditional markets, including power stations, aluminium smelters and data storage and analysis.

Personnel

Case M6 consists of two sites. This one had 300 employees, is concerned with production, and is situated in Lancashire over 150 miles away. The geographical separation between the two is indicative of the way the company wishes to keep design and production separate. The first site is for conception, design, drawing, the building of prototypes and fault-finding. Once this process is complete, the final specifications for manufacture are sent up to Lancashire. The object is for as little discretion to be left to production as possible. One of the engineering managers put it thus:

> there are too many clever people about who insist on thinking for themselves. You often get the shop-floor putting things right on a unit without telling anyone, so that the factory will carry on making the same mistake. This is serious when the factory is that far away.

This was referring more specifically to the blue-collar workers working on the prototypes in the design factory, but the principle is the same. The more complex the technology, the more the management is likely to insist that all minor modifications pass through the control of engineers and engineering management. There is an analogy between this and the tug-of-war between the drawing office and the machine shop in Range-finders. The pressure (in which strategic choice plays as much a part as the technology) is for data to be collected, analysed and adapted at a series of levels, the lowest of which is clearly identified.

The company over the two sites has 640 employees – there are thus around 300 at the production plant. The number of employees has declined at both, although precise figures were not available either for that or for the fluctuations of turnover. It appeared that the company had passed through a period of relatively poor performance, perhaps due to stagnation in some of its markets, such as ship-building and coal-mining. A new managing director had been appointed, under whose auspices an almost defunct personnel function was to be used to play a significant part in bureaucratising the 'family firm' culture and emphasising the high skill-intensity strategy chosen.

Employment on the design site was as follows:

Commercial, sales/marketing, customer service	60
Engineering	185
Personnel, Finance, Administration	20
Blue collar	75

Engineering itself was divided into 125 engineers formerly working in product divisions and now organised on a project basis, and fifty-five in engineering services. In addition, there were ten engineering indirects. Engineering services include CAD, testing and small scale manufacturing.

Skills and training

As at L6, engineers can follow either a managerial or a technical career path, with the technical consultants being viewed as boffins who have little practical sense. Systems expertise is required more and more as, increasingly, standard components are placed in more complex arrangements which can be adjusted for a variety of applications. There is at the same time a hostility to software specialists – some of these, who reputedly 'didn't know one end of the product from the other', had been 'farmed out'. People are being recruited who are hardware-based but who have training in software on their degree courses. Almost all those in engineering are graduates. Test engineers will be HNC and some of those in the drawing office will be chartered engineers. In connection with hardware/software, we should say that in some aerospace applications, such as military helicopters, systems are microprocessor based but with an analogue override. This trend will prevent a wholesale skill shift from occurring. At the same time systems generally are said to occupy one-third of the space but have three times the complexity in terms of functions. To enable the effective development of 'firmware' specialists the company has instituted an instruction course of two years based on an Institution of Electrical Engineers model, for teaching hardware to software engineers. For the first five years, engineers are expected to be able to operate in both. Later, they can specialise.

Recently, initiatives from the new managing director had been made to set up a training programme based on appraisals in order to find out what was required in engineering and how best to achieve it. It was felt above all that engineers needed a greater level of company and product awareness, as well as managerial skills. This programme under a new personnel manager and personnel officer, was still in its early stages. The training budget is still limited, however, and the Personnel Department hope to use the training needs appraisal as a lever to get more money and in the process increase their importance. New technology (taught through suppliers' courses) is one of a number of initiatives already embarked on. Links are being made with polytechnics in order to have courses on systems engineering. Courses are also being negotiated to enable a switch from hardware to software or vice versa. Four graduates are recruited a year. These are considered to be at risk through poaching by larger companies for two to seven years. Employee turnover is 10 per cent overall. Salaries are thus maintained at the bottom of the top quartile of the parameters agreed amongst electronics employers. The obsolescence of ageing engineers (over 35) is seen as inevitable and is seen to be a problem that will eventually go away of its own accord. The training budget, at £30,000 cannot achieve much to change this as yet. The technical line (for specialists with no overall systems knowledge), is seen as a place into which boffins who have become obsolescent may be housed harmlessly. For the last few years, only electronics graduates have been recruited.

M7: 'STABILIZERS LTD'

General

Company M7 were leaders in mechanical engineering devices for shipping and the offshore oil industry, notably stabilisers, anti-heeling devices and motion control generally. They have been very seriously hit by the shipbuilding recession and, then, the collapse in oil prices that froze North Sea oil development projects. As a company, they belong to a larger group but as the marine division they have considerable operating independence. The arrival of microelectronics over the last few years has led them to set up an independent company under the same group umbrella and situated across the road.

Personnel

The turnover in 1987 was the equivalent of £24m of which £20m was derived from sales, the rest from customer services.

The breakdown of the company was as follows:

Shop-floor	
Unskilled	40
Skilled	160
Foremen	14
'Technicians'	70

Including draughtsmen, sales, purchasing, ONCs and HNDs.

Full List

Administration and professional (Accountants etc.)	11
Clerical	36
Office supervisors	4
Foremen and other supervisors	22
Mechanical engineering craftsmen	161
Pattern makers	2
Toolroom fitter	1
Maintenance engineers	16
Welders	21
Electricians and plumbers	9
Semi-skilled or unskilled labour	38
Apprentices (already included under trades headings)	(40)
Management (including four directors)	22
Scientists, metallurgists, etc.	8
Total	351

The scientists (and some of the management) are all graduates. Around twenty-five in the technician belt have HNDs, notably the engineering draughtsmen.

The cutback, as at L5, was across the board. Although the aim was to reduce shop-floor labour, retirement and redundancy were widely available, with many obsolescent managers leaving. As elsewhere we were told that an engineer's best work was done by the age of 35 after which he should find an administrative role or leave. The company was trying to shake out of its old family image. As at M6, one measure used in this attempt was an expanded training programme under the leadership of an engineering manager of high status nearing retirement (of Alarms Ltd).

Skills and training

The training budget had been increased from £100,000 to £160,000, and was to be raised to £200,000 subsequently. This should be taken in connection with a turnover on a reasonable year of £20m. A large part of the budget was taken up by the high levels spent on the apprenticeship scheme (the company clearly did not equate reduction of employment with a freeze on recruitment and training) and management training.

Other training concerned word-processing, sales, CAD and CNC. Further training included the sponsorship of two people on Open Tech courses and four on the Open University. The latter course concerned electronics appreciation.

As far as graduates are concerned there were twelve, including those in management. In addition, there were four higher degrees, three of which again were in management. On top of the twenty-six HNC/HNDs, there were twenty-six ONCs.

Cost-cutting in manpower and in time was assisted by investment in CAD and CNC. CAD enabled the utilisation of standard components and designs, where previously everything had been done from scratch. The company now had no employees over the age of 60. This decision was in line with the Group policy of retirement at 50 if desired, with a more than 50 per cent pension. The training manager claimed that the retirement of so many of the long-service employees had been assisted by low inflation and high interest rates in the national economy, which meant that retirement and the associated investment of lump sums carried a much smaller risk than in the past when people had struggled on as long as possible out of fear of an impoverished retirement.

Chapter 7

Case studies: small sites

S1: 'CIRCUITS LTD'

General

Company S1 was a producer of small-scale printed-circuit plating lines. The market for such lines was small, and a 50 per cent share of the domestic market plus exports had to be maintained to ensure profitability. The small-scale producers who bought the equipment had, in turn, to maintain a high level of flexible utilisation of the equipment to compete with larger producers who could offer economies of scale. This in turn put pressure on the firm to maximise the flexibility of the systems they constructed. Over the last four years, the company had developed a software-driven system. This system, which was based on an earlier mass-production model but modified to give much greater flexibility, had increased its share of company sales from 15 per cent to 85 per cent.

Despite this success, the company preferred to remain small. Expansion from £1m annual turnover to £5m was seen as possible but undesirable in terms of extra administrative effort, the recruitment that would be required and the general fear of diminishing returns.

One way in which the company attempted to maintain or increase its market share without substantial increases in overheads was represented by the increased 'factoring' of the product, buying in hardware, either mechanical or microelectronic in an increasingly finished form, and contracting out whole sections of the production process (as in the case of ironwork). Significantly, this was not done in the case of software development. The company was seen as expanding in the direction of more research and development to produce highly sophisticated control equipment. This would present a problem in terms of skills expansion rather than overall expansion. Already the company could be described as skill-based rather than technology based. In future, it might require a greater concentration of software expertise.

The customers' demand that was being met by the company was for systems with a high level of direct management control and flexibility. The investment spent by customers (generally smallish ones on such equipment) was of such an order that maximum utilisation was desired in order to get adequate return on capital. Thus, systems provided for monitoring (or even operation) direct from management offices. Recent models were able to provide a store of retrospective data on the system's operation for break-down analysis, so that not only were management informed immediately of any delay that occurred, but they were also able to determine what had caused it. A major 'leap forward' had occurred in the programming of the system over the past two years. It was now possible for the operator merely to press buttons on the outside of the control box to state what was required for the system itself to choose what currents were necessary and set them in motion. It was also possible for anything up to a week's production schedule to be entered at any one time (presumably by an official), leaving the operator merely to mind the system and use the manual override if necessary.

Personnel

The company had long stabilised at twenty-five employees:

Directors/senior management	2
Management	3
Administration	3
Sales	1
White-collar engineering	2
Electricians	3
Fitters	3
Other shop-floor	8

Skills and training

The individuality of customer demands (which necessitate a type of 'consultative selling' – and thus considerable understanding of the systems by the involved, namely the sales manager and the managing director) prevented the work of the company's programming (and to some extent that of the three or four directly involved in electronic assembly) becoming routine. It also meant that the company could guard its skill/software resource of programme generation, to which customers were obliged to return to the company. Whilst the company could not afford to mount any elaborate field service (which might have been subject to de-skilling given the simplicity of the new assemblies) it was able to make programming skill and facilitate its main after-sales commodity.

Microelectronics had had relatively little effect on the production process and the skills involved, at least as far as the shop-floor was concerned. None the less, the overall building time for a complete system had virtually doubled. This may have been due in part to increasingly specific customer demands. Certainly, the planning and designing functions took up more time. There was a high level of skill required from the electronic assembly workers, according to the directors. According to the programmer (who had been promoted from the electronics group) the work consisted of straightforward assembly. When electro-mechanical overrides were done away with (this had not yet occurred) a decrease in skill level might be expected. Indeed, the tendency to increasingly import complete electronic assemblies would work against higher skill content in these jobs.

Up-skilling had of course occurred with the promotion of one of the electrical workers to programmer. Although he had received some computer training as part of an HND course some years earlier, he was generally self-taught. Microelectronic knowledge elsewhere in the company (notably among senior management) was also self-acquired. No status difference appeared to exist between the programmer and workers who had no such expertise. The lack of precise abilities in this area on the part of superiors in the company gave the programmer more autonomy and higher (informal) status than the same function might secure in a larger concern. The viability of customer demands and the need to find an optimal fit between these and the fixed constraints of the chemical process appeared to guarantee the non-routine nature of the job.

As we have said, some skill shortages had been encountered regarding shop-floor jobs. This was not seen as being connected with microelectronics, and had largely been solved before the expansion of microelectronic involvement in the product. It was more due to the lack of industry in the area. Workers of all types tended to be recruited locally, and taken straight from school where possible, often receiving their training informally on the shop-floor as a result. Once employed they were not trained or retrained in any outside institution.

S2: 'PACKAGES LTD'

General

Very small in size, but well-known in its sector, Company S2 was involved in the design, marketing, customisation and installation of CAD packages. Founded on strong academic knowledge, it failed through cash-flow difficulties and went into receivership late in 1985. The fact that the receiver was said to be a specialist in winding up CAD companies suggested that this kind of failure was a far from isolated occurrence.

The firm was sold to a local electronics group for a surprisingly low price, which appeared to disregard the value of its laboriously-cultivated user-base. One of the pitfalls of the firm's business strategy was its reliance on one large supplier for the hardware used, to which S2 added the software and services. Senior personnel in the company had debated the idea of selling software alone, but this move was put forward too late. Sharp competition in the CAD systems market had required them to provide a total package involving hardware, software and extensive services.

Personnel

The company had eleven employees:

Executive directors	3
Systems support manager	1
Software creation	3
Technical support	2
Sales	2 (+ 3 vacancies)

Skills and training

All the employees had university degrees or the equivalent, with higher degrees among the directors. Of the software engineers, two had non-software degrees and had taken postgraduate conversion courses.

The company's problems in this area arose with the sales staff, of whom three had been poached in the weeks prior to our visit. The company's strategy required a broad range of skills from its sales staff. These have to both technically well-qualified and still able to sell well. Unlike such companies as M1, where it was seen as sufficient to give sales staff a week's training prior to the launch of a new product line, in the complex and problematic area of CAD systems, where installation and its attendant problems may take literally years of attention, the company found it difficult to attract and keep suitable broadly skilled sales staff.

Selling and installing the company's systems involved not only highly detailed negotiation and modification of the product to customer's requirements (which in the highly competitive market grew increasingly detailed), it also involved training the user company's staff. Whilst sales staff of the right technical level would find the negotiation rewarding, they were inclined to become bored when it came to training novices. The small size of the company seemed to require that both jobs be done by the same people.

Company S2's experience contrasts with that of S1. Whilst S1 also involved itself in 'consultative selling', it refrained from offering any post-delivery services or training to customers on account of its small size. In the case of S2, it had developed links with educational institutions for the running of courses of up to four months' duration for customers. Thus, training represented another way in which the company was 'exploited' by its user base. It was ironic that, after such extensive cultivation, the user base was not regarded as a company asset. It did however (as was the case with S6) provide a means of recruiting highly skilled personnel, such as the systems support manager. Like many such companies, S2 provided a very congenial working environment and varied tasks. These benefits may not have survived the take-over by the larger concern, which was shortly followed by the (unsolicited) departure of key personnel.

S3: 'DEVICES LTD'

General

Case S3 was a small mechanical engineering firm operating within a larger concern. It had been badly hit by the contraction of the automative industry in the region and had been pushed to the point where it had been expected to close down. Instead, new top management had been appointed to cut costs, redirect the product strategy, and turn the company around.

The company had, for its small size, a bewildering number (over 100) of 'tried and tested' mechanical product ranges, most of which had lost their main markets. The strategy was now to instigate a micro-electronics-based product range which would gradually replace the old lines. The target was for these new products to take over 50 per cent of the firm's business within ten years. The process had been started six months earlier and the first products were beginning to appear, although they as yet accounted for only 1 per cent of business.

The project was under the direction of a new deputy engineering manager who was involved both in developing the new products and in recruiting the necessary staff.

Personnel

The total employed in the firm was 180:

Management, Administration	30
Engineering and engineering management	16
Sales and marketing	22
Direct production	110

Prior to the recent cost-cutting drive, however, there had been 240 – the losses were mostly in direct production.

The electronics complement so far included two electronics graduates (including the deputy engineering manager) and one technician (ONC). Two electronics assemblers had been trained from the shop-floor. The target was four graduate engineers and one technical.

Some difficulty was being experienced in filling these two positions, partly on account of the low pay offered. Apparently this was being changed and electronics engineers would be receiving higher salaries than their grade would have given them before. It should be pointed out here that two contradictory traditions influenced the company – one former parent company was, and remained, a relatively high-tech company, whilst the current one had a tradition of being under-capitalised. As far as shop-floor skills were concerned, there was no difficulty in finding people for the new jobs involved. At the start, only one assembler was required and it was simply a question of the project manager asking line supervisors (whose workers were mostly women and involved in semi-skilled manual assembly work) who would be most suitable to learn new skills. There were now two women working on PCB assembly, while for the rest of the process, workers and facilities from the mechanical side were borrowed.

As far as filling vacancies for electronic engineers is concerned, difficulties would be likely to result not only from the shortage of suitable applicants, or the relatively low pay scales, but also from the fact that a company such as S3 using single-chip processors in a fixed control function would be unlikely to provide an appealing career structure for an electronics engineer. Discussions with managers from other small mechanically based companies who had difficulty filling places for this reason, suggested that there was no easy answer to this for companies forced to incorporate a marginal amount of microelectronics in their products, except to train mechanical engineers to be able to write software, assess circuit designs, etc.

Skills and training

The company provides a good example of a mechanical engineering company in the earliest stages of switching to products involving microprocessors. At all levels small adaptations, through formal or informal training, were having to be made, although the main shift – the establishment of an electronic design department – would be effected through the hiring of already qualified graduates. Further expansion may have to rely on training engineers from the mechanical discipline to work with electronics.

The inclusion of microelectronics was seen to have less impact on the shop-floor. Some detail was given on the shop-floor skills required for PCB assembly and testing. Four stages were involved in the process.

1 Production layout. This is taken care of by the machine shop as for mechanical product assembly. No extra training is considered necessary.
2 Chip soldering via surface mounting. The technique of surface mounting is used. This was considered to be the most skill-intensive part of the shop-floor process. Although surface-mounting involves less manual dexterity than traditional techniques, it requires greater placement skill and is thus viewed as more demanding. Machines exist for automating this part of the process even for the small batches which would be required here. The cost was, however, seen as prohibitive.

   The company had not yet decided whether to send the assemblers on a component recognition course at a local college. Otherwise they were basically self-taught on the job. This does of course slow production as the learning curve proceeds. One problem is the thought-repetitiveness of this stage. The placing of the components, while requiring less dexterity as said above, requires greater attention and thought. The engineer considered that 'thought' monotony was more harmful than 'dexterity' monotony. The latter could be carried out at speed while the assembler's mind wandered wherever it wished. With the former it was essential to concentrate continually without the stimulus of variety in the problems to be addressed.

3 Testing. This is relatively simple – the PCB is linked to a computer which registers a signal of 'pass' or 'fail'.
4 Spray and final assembly. This, again, is relatively straightforward.

At the pre-production stage more complex testing and diagnosis is carried out. The skill here lies in thinking of all the ways in which a PCB may reasonably be upset and subject it to them. An earlier stage still is the comparison of circuit diagrams by the project engineer, who determines which is the most suitable for the intended uses.

The electronics project engineer came from an electrical background which was essential for his role in liaising between marketing and design and production – that is, his ability to relate circuit diagrams to product use could not be acquired any other way. It was also necessary for him to deal directly with the research department at the other site.

The programming would not be in a computer language as such, but in a machine code. This would presumably make it relatively easy for non-software trained engineers dealing in simple single chip control equipment to programme. It should be mentioned that no actual design of the electronic component of the product is carried out at the site, although it will be in future. Rather, it is the decisions on the design that are taken there.

The importance of design, assembly and testing skill in the company in future is likely to be vital, since, as a low-volume producer, with FMS only a few years away, they have no significant price advantage and just sell on the quality of the product. This also meant a much greater emphasis on customization, which in turn meant greater co-operation between Engineering and Sales, and extra training for the latter.

Liaison is increasingly important between the two departments in that the sales people have to be aware of the cost of any deal they make. Engineers are thus being made more cost-conscious and sales and marketing personnel more 'engineering conscious'. Training is to be applied for sales people to understand the technical advantages of the new products. In programming the chips, engineers attempt to leave space where possible to add on new functions (such as clocks) to appeal to the potential customer. If, however, the marketing department tell them that only the tachograph (for example) is required, then they will use all the programming space to refine and improve the basic function of the instrument.

Training is necessary not merely because the products are going to change more often than before and in a new area, but because sales staff will have to understand the rudiments of electronics design since they will have to explain to the customers the various options available within the range and the ways they may be customised. More than mere descriptive knowledge will be necessary, but rather a grasp of the principles and general capabilities beyond the scope of any one product.

S4: 'TESTING LTD'

General

Company S5 had been formed in the early 1980s as the manufacturing off-shoot of an American-owned distributing group. It was involved in small-batch precision engineering. Pressure of deadlines as the company's two product lines were launched led to what was seen as an overdependence on sub-contractors and less emphasis on training.

Personnel

The firm had thirty-three employees:

Managing director	
Development technicians	5
Development fitter	1
Production control/supervision	6
Assembly fitters (mechanical)	6
Assembly and test (electrical)	3
Semi-skilled electrical assembly	6
Stores	2
Draughting	1
Metal cutting	2

Skills and training

As far as skilled/semi-skilled proportions are concerned, the employees may all be counted as skilled, with the exception of the six semi-skilled electrical assemblers, the two stores workers and the two metal-cutters. One of the development engineers graduated after six months with the company (one year previously). No other graduates had so far been employed. There had been no turnover of personnel.

The company policy seemed to have been to take on anyone available and worry about long-term suitability later. The priority was to have the units produced at all costs, and personnel shortages were not permitted to hold this back.

Software provided an area of potential difficulty, although this was easily enough overcome in that the company hired for its development team a self-taught programmer who was known to be adequate since he had previously done sub-contracting work for the company. As well as advertisements in local papers, this informal method of recruitment should not be underestimated. Now that the order backlog had been dealt with the divisional manager was considering more of the strategies practised by larger and more settled companies – more systematic recruitment, training of existing personnel on outside courses, links with local schools and colleges, etc.

One specific area of difficulty had been the company's wish to recruit a mechanical design engineer who knew and had experience of the electronics industry, preferably with an understanding of semi-conductors. Nobody would reply to advertisements asking for such a person. A gap was perceived in the education system in that it provided engineers too specialised in their particular field, whilst industry required mechanical/electrical hybrid qualifications. This was especially relevant in that the company were providing mechanically-based support equipment for the semi-conductor industry.

Some jobs were put out to contract or contracted in. Printed circuit board jobs tend to be contracted out, although all development and prototype building takes place within the plant. Some of the manufacturing final assembly has occasionally been put out to contract.

Production planning (geared to sales forecasts) was seen as vulnerable to skill shortages and a resultant over-dependence on sub-contractors who at this stage still account for 80 per cent of the production process all told. To produce the microtesters, a lead-time of twelve weeks is required, whilst four weeks' delivery is apparently essential to competitiveness. At present, the company were attempting to reduce their inventory (they had previously been behind on meeting orders).

As far as skills in development work are concerned (as opposed to the actual building of prototypes, for which most of the necessary people and equipment existed – metal-cutting, wiring, etc.) there were shortages as far as formal qualifications were concerned. Both of the development engineers/technicians who write the software for the microtesters were self-taught, one of whom also had hardware expertise, the other having been a sub-contractor for the company as already mentioned. One other had electronics knowledge, the other being on the mechanical side. The company appears to derive its surplus value far more from its mechanical design and precision than from its software, hence its lack of immediate concern over lack of software expertise. As the divisional manager said, they were not blind to software but they were primarily a mechanical engineering company, with, as said before, a need for electronics knowledge resulting not from production requirements but from customers' requirements – in other words design was rooted in mechanical engineering but had to be attuned to the needs of the electronics/semi-conductor industry.

Due to production pressures, training as such had received little attention in S4. Only now was the possibility of developing the workforce being considered. Links with local higher education colleges were also being investigated.

S5: 'VIDEO LTD'

General

Company S5 was situated in a stable and growing market for video-related products. The inclusion of microelectronics in its products had largely meant the replacement of existing functions rather than any radical departure. Since all assembly was still carried out entirely on the site, the shop-floor had not been drastically affected, even though complex electro-mechanical assembly was no longer required. The company had recently been taken over by a larger group. Despite a much resented increase in paperwork, this had not affected strategies regarding products or personnel. Regarding the latter, the company had evolved a means of temporarily relieving potentially serious skill shortages through the offer of training for recruits.

Personnel

Working directors	2
Despatch	2
Production control	2
Data processor	1
Sales	4

Sales administration	2
Accounts	2
Research and Development (plus one vacancy for prototype wireman)	3
Mechanical assembly	5
Sprayer	1
PCB assembly	3
Electrical assembly	6
Quality control	2
Specialist electricians	3
Total	39

Skills and training

A recruitment problem regarding specialist electricians and electrical engineers is a feature of the company, which they have found a way of coping with. Otherwise problems are experienced in the area of skilled manual labour – as in the case of prototype wiremen. This difficulty was put down to the regional labour market in that the traditional industries in the area had been furniture and paper. Financial and career considerations tended to mean that no one with either draft skill or formal qualifications would stay long in the company (except for those who had 'grown up' with the company and occupied senior positions. The more unskilled workers tend to remain with the company. The average age of the employees thus rises with the company. Most on the production side are in their late 20s. As these become settled, turnover is expected to decrease. By and large, turnover is around 10 per cent overall. Like other companies of similar size, they appeared loth to expand employment too far even though orders were outpacing production capabilities. A degree of buffering was provided by overtime flexibility. The need for overtime, it was hoped, would be diminished once the stock and order processing had been computerised.

No formal training was conducted within the company. On the other hand, skilled engineers could not be acquired 'off the shelf' in the area, since it had few traditions of capital-intensive industry. School leavers from the two grammar schools in the area moved, 'usually via university', to the larger companies with their well-defined career structures and offers of more interesting work. In practice, the company found that school-leavers from the comprehensive schools, if they were of the right standard, were on the whole more self-motivated. All three of the specialist electricians had come from comprehensives straight to the company and were the only group of trainees in the company at that time, although unskilled workers in the company could be counted as trainees while they received their on-the-job training. As explained above there would be fewer and fewer of these as turnover in unskilled personnel declined. The practice for specialist electricians was to recruit them straight from school, on the condition that they should work towards an HNC qualification during the time they worked for the company. This they would do on day-release. At present there were three on an electrical HNC course at the local College of Higher Education and two on a mechanical HNC course. The course would take five years on one day a week release. One result of the increasing shift to microchip technology was the fact that eight or nine engineers, including several mechanical, used to take the course from the company three or four years ago. Having received their HNC, engineers tended to leave, usually within a year, at the age of 21 to 22. This was unfortunate since, it was said, the job involved a five-year learning curve.

The specialist electricians carried out important work in that they developed prototypes and customised standard products. Given that the company's labour costs approached 20 per cent of turnover already, they did not consider that they could afford the top market rate so as to avoid the need for trainees.

Higher up, the skill requirements seemed informally determined. One of the few graduates in the company (possibly the only one) had been chief engineer for six years, and had now been promoted to technical director. He was succeeded by a specialist electrician who had done his HNC and stayed on. He took over the same job in research and development but not the title Chief Engineer. This description was assumed to refer to a graduate, which he was not, and it was no longer considered necessary to have a graduate in that job.

Recruitment was also informal. There was no Personnel Department or official. Recruitment was carried out throughout the year on an *ad hoc* basis, with the chief accountant taking a major part in selection. The technical director had a degree in design engineering but had broadened his scope and had taught himself programming etc.

S6: 'ANALYSERS LTD'

General

Case S6 was an expanding high-tech small-batch company, designing and manufacturing analysers for an increasingly commercial market – it had formerly served a primarily academic range of customers. This shift may be accompanied by some cultural change, since many of its key staff had been recruited from the existing user-base. The company has expanded recently through the acquisition of a US subsidiary. It has increased its shop-floor employment in a drive to reduce dependence on local sub-contractors.

Personnel

The company increased its workforce from 120 to 180 by the mid-1980s. This growth was connected with the launching of new products over the next three to four years. The product cycle is accelerating – although systems from twelve years ago are still working in many cases (usually in universities) their lifetime is five years or less. This requires a disproportionate investment in R&D – which accounts for 30 per cent of the turnover and around half of the personnel. Many of the new recruits were graduates, who together with HNCs (with whom they are lumped together) make up 60 per cent of the workforce, most of the remainder being manual workforce, usually semi-skilled.

Management and administration were kept down to the very minimum. The breakdown was as follows:

Directors (all working)	4
Management	10
Administration	6/20 (four accounts, two secretaries)

Changes had recently occurred at the board level. Out of seven members, five had been purged, of which two were replaced. The result was a reorientation towards marketing and away from pure science.

Direct production (mostly semi-skilled) has been incrreased from twenty to fifty in a year, in order to meet deadlines which sub-contractors (on PCB assembly etc.) had failed to meet.

Delivery of products is usually targeted for six weeks. The general manager found sub-contractors spoiled any attempts at planning. As far as chip-manufacturers were concerned, these were considered so unreliable that a year's supply was always kept in stock.

Research, Development and Design covered about 100 people of whom ten were Ph.Ds, and eighty were graduates or HNCs.

Skills and training

The company claimed to have had few problems recruiting suitable staff. They attribute their success in this regard to the fact that large numbers of university students in the relevant disciplines (for example, chemistry), were already familiar with the company name, having used its equipment. This fact enabled the company to recruit people that would normally have been snapped up by larger companies on the 'milk round' (in which they did not participate). Thus, for the most specialist posts the company had a ready market and, through university departments (its traditional customer base) an informal recruitment network. Whilst chemistry departments were perhaps the most frequent academic customers, the flexibility of such high-tech products meant that it had found niches in many other scientific disciplines, thus making recruitment easier in them too.

As far as software engineers were concerned, these were 'not liked' by the company, who preferred to have graduates from other disciplines who had learned software rather than the other way round. In practice, the disciplines – mechanical, electronic, chemical, physics – overlapped on account of the complex nature of the product and its wide range of applications. Overlap was necessary since the systems tended to be customised for each buyer (hence their being designed on a strictly modular basis). Some knowledge of the discipline into which a given system or group of systems was to be applied was considered necessary – presumably it would be acquired informally on the job, a process of adaption that would be easier for natural science specialists than for software engineers. None the less, software has a key role in the product, particularly since the company actually produce their own computer as part of the product.

The company has moved against the trend of going to sub-contractors on account of skill shortages. It was in fact the unreliability of sub-contractors that led them to increase their staffing, particularly at the blue collar level, whilst the graduate sector expanded due to an upgrading and decommercialisation of the product range. The aim was to be less dependent on the electron microscope and to build up other areas such as material analysis. For a company of small size they were becoming caught up in the familiar and potentially dangerous cycle in which as soon as a technique is mastered it is miniaturised so that the whole of the product (as opposed to its constituent parts) becomes more complex with functions enclosed in a small space. This change has had several effects which have in their turn influenced its HRM and employment policies:

1 Delays due to sub-contractors are arguably more serious in the sense that the number of components and sub-assemblies becomes greater and their strategic importance increases. This is of course one of the areas where a paradoxical movement occurs, in the sense that on the one hand the number of components in a unit that are individually bought decreases, whilst on the other the number of 'units' increases. Much of the confusion arises on account of changing definitions of what a 'unit' is and what it can perform, a question to which we shall return but which we cannot go into here. Put simply in this case, PCB assembly may involve fewer boards per unit, but delay in receiving chips for a board will have a graver adverse effect than before. Equally for PCBs – there are fewer involved but each contains more functions and is more complex than before. In the case of S6, delays in this strategic area are now more serious since they are competing in a commercial rather than an academic market, it being reasonable to suppose that deadlines and the risk of losing contracts becomes proportionately more serious.
2 More positively, as far as the company is concerned, these technical developments (multi-layer boards which have computing power of twenty megabytes rather than four, as was formerly the case) have facilitated customisation, the product consisting of a number of building blocks. Clearly the more functions that may be contained in a small space the greater the flexibility of application. This again goes well with the company strategy of reaching a wide range of firms in the commercial sector, with different people wanting a different type of microscope etc. Much effort has to go into compatibility with users' equipment. There are 300 mechanical interfaces which come before a signal is given. This has been instrumental in the company's move to stand-alone systems and the development of their own computer. This has meant image enhancement in analysis and better results generally. Until this becomes the norm, the company has to provide mechanical piece-parts to enable the systems to interface with user's equipment.
3 The two developments that have resulted in stand-alone systems on a modular basis have enabled the company to increase its batch size accordingly from ten to fifty. Software naturally plays a large part in customisation and flexibility of application. As all software is carried out in-house, there is a weekly software and application meeting

(these meetings play a disproportionately important role in the company given the very slight amount of direct supervision carried out).

These developments, as we have described, led to the company deciding to carry out a large amount of PCB assembly, normally subcontracted, inside the factory. To complete the cycle described above, the increase in batch-size meant a greater emphasis on planned production which again was too vulnerable to sabotage through subcontractor delays. Accordingly, the company set about hiring assembly workers. Paradoxically they found it more difficult to find experienced workers with the right skills in this area than they did to find suitable graduates. PCB assemblers are used by a number of companies in the immediate vicinity whilst not being part of the traditional labour force in the district (a problem similar to that found by S5, another local company). Like company S5, S6 went to get its skilled manual workers from the available school leavers, of whom there were no shortages. These were then trained in the company which runs what it calls an EITB-style system, although it was admitted that 'they don't come up to Engineering Industry Training Board (EITB) standards, but they serve us well'. Training is clearly seen as a growth area, the company being on the point of appointing a training manager, a significant move in a company so loth to increase the number of its indirect staff. The EITB-style scheme was however not a new development.

Viewed cynically, the recruitment and training strategy has the advantage of taking on the youngest workers with no skill bargaining-power and giving them training which goes far enough to meet the specific demands of the company but not far enough to give them an edge in the local labour market. The manager said, almost in so many words, that while he might be biased, it seemed as though qualifications meant increased turnover of staff. Notwithstanding, the company did, where it saw fit, sponsor people in formal courses via day release, whether four years on the HNC or six years on the B.Tech. Six electricians were also on a computer-servicing course, presumably with a view to being deployed for customer service. It seemed implicit that the lack of an appropriate career structure meant that most people would leave on gaining their formal qualifications. Higher up, however, little problem was seen as far as keeping graduates or postgraduates in the company was concerned. These were usually chosen for their ability to be self-motivated and committed to their R&D work, Ph.Ds in particular were regarded as people who were quite happy to be put away on a job and left to get on with it. They were not expected to have ambitions that might be thwarted by the lack of career progressions available. This characteristic was very likely true of the chemists etc. as well, in contrast to the software engineers that the company so disliked. The company were thus exploiting habits and attitudes that belonged with their traditional academic market by applying them in a commercial context. The software engineers who are more close to that commercial context than, say, chemists, could provide a destabilising influence, through attitudes as much as through the differentials they might demand (as had proved the case at Company M1, where the 'happy plodder' view of engineers had receded under this outside influence).

Another way in which training and technology had affected each other lay in training at the higher levels (it was stressed that training took place at all levels). Apart from conventional graduate induction, this has involved some cross-disciplinary awareness between mechanical, electronics, chemical and software engineers, to go with the hybrid nature of the product and the complexity and diversity of its forms and applications.

S7: 'SYSTEMS LTD'

General

Company S7 specialised in designing conference systems, 90 per cent of which were for export. The company was founded by a nucleus of people who left a long-established company in the audio sector. The numbers grew slowly, reaching twenty at the time of interview. The company was wary of taking on too many people on account of its uncertain cash-flow and its dependence on a small number of large contracts each year. At the same time, difficulties regarding the recruitment of skilled personnel represented the most serious constraint on its expansion. One way in which it was seeking to stabilise its position was through the launch of a range of more standardised 'bread-and-butter' products in the audio range. Individually, these units represented the smallest fraction of the cost of the larger systems but would provide a more consistent source of income. In all, they accounted for 10 per cent of turnover, although they were planned to reach 50 per cent in the foreseeable future. This range of smaller products did not as yet involve microprocessors; the main product range, that of elaborate conference systems had, on the other hand, been revolutionised through the application of these over the last few years.

Were the newer simpler products to establish a footing, the company would be tempted to move more fully into manufacturing, most of which, at present, was contracted out. Thus, not only would these new lines lead to a more constant cash-flow, they would also ease the company's reliance on sub-contractors. Uncertainties regarding these had on several occasions had adverse effects on lead-times. This was potentially serious given that competition in this area of business was more concerned with delivery-dates than with price.

The other strategy open to the company is to adapt its conference system expertise for other applications, such as nurse-call systems. This would be concomitant with the firm's becoming more a 'systems house' than a manufacturing unit. This strategy, and the role for the company that it embodied, was advancing gradually, but was being severely constricted by skill shortages, as were projected moves into new technical areas such as microwaves.

In the meantime, the conference area is developing rapidly. On the one hand, the possibilities opened up through the use of large numbers of microprocessors is immense – one speciality is simultaneous translation whereby hundreds of channels process and transmit data with minimum cabling or power supply. On the other, the customers' awareness of these capabilities, to which Case L7 has contributed very significantly, has resulted in a raising of expectations that is difficult to meet – 'technical sophistication has increased requirements – you could never get away with now what you could have five years ago'.

Personnel

The company employed twenty-two people:

Managing director	1
Technical director	1
Administrators	3
Production supervisor	1
Systems engineer	1
Firmware engineer	1
Draughtsman	1
Production (wiring and assembly)	5
Prototype wiring	5
Test engineer	1
Trainees (HNC)	2

Skills and training

Of the personnel given above, only four were graduates (the two directors and the systems and firmware engineers). The key figures were still those who had left the larger company previously. As would be expected in a small firm geared to producing complex small batches for large, highly customised contracts, the demarcations are relatively flexible. All staff are monthly paid.

The company has a policy of developing the skills of people already in the company on a long-term basis. Most of the employees are over 30 and unlikely to leave the company. Only three people have actually left the company since its foundation – one because she didn't wish to commute, another dismissed after two weeks because he was found not to possess the qualifications or skills he claimed to have. The third was a systems engineer who left for a much higher salary. The company recognised that it had made a mistake in this respect, since many engineering firms in the region were looking for just that type of person. Since then they review pay levels every six months and make certain they are offering the market rate – 'you always need to be looking over your shoulder nowadays'.

When staff are lost, they are not easily replaced. It took four months before a new systems engineer was recruited. It had taken a similar amount of time to recruit the firmware engineer and six months to find a production supervisor. Printed circuit board assemblers and prototype wiremen have also taken several months to find. The company had received some publicity in the national press when they declared themselves to be on the point of closing down on account of skill shortages. They have been inundated with applications since then. Unfortunately, almost none have had the right qualifications or experience. Although the company were exaggerating about closing down, there was no doubt that they were increasingly unable to expand in line with their technical and market potential due to a lack of suitable staff.

They were currently trying to find another engineer qualified in software. So far, they were still only designing a small percentage of an expanding software component in the product and were feeling squeezed by the high rates charged by the software houses on whom they were dependent for the rest. Having decided that it would be cheaper to employ a second one, they encountered severe difficulties. Being situated in the south-east of England, they were competing against large firms who took all the available computer science graduates via the 'milk round'. They did, however, find – through nationally placed advertisements – that there were suitable graduates in the north-east of England. As we found with several other companies which tried to recruit from this source, no package that could be offered would compensate for the huge differential in house-prices between the two regions.

Formal training played only a small part in the company. The need to develop software expertise had been met at first by the technical director teaching himself in his spare time until suitable people could be hired. The engineers supplemented their knowledge in this way, but were unable to be spared for courses in outside institutions. Training for other employees was also on the job, where they were expected to fulfil tasks beyond their formal responsibilities in the firm. The company had, however, taken on two unskilled school-leavers as trainees earlier. These were being sponsored on a four-year City and Guilds course via day release. The course dealt with a wide variety of skills associated with printed circuit boards and electronics generally and it was hoped that they would play an increasingly important part in the company if they stayed beyond graduation (this was by no means certain, as we say with S5). Since manpower was stretched, trainees were expected to put in extra hours to compensate for their day out.

A tendency towards hybrid skills was more noticeable in this company than in many of the others. Some cases of this were straightforward – the draughtsman adding electronics draughting to mechanical draughting for example, the production supervisor and his subordinates dealing directly with customers and intermediaries. More particularly, there were the examples of the systems engineer and the technical director, and to some extent the wirepersons. All these were involved in hybridisation on two counts:

1 The business was almost entirely customised by definition, every conference provided for being in a different location and having different needs. The systems engineer was involved in negotiating the details with customers. Although not expert regarding the internal working of the equipment, the systems engineer had to arrange modifications regarding colour, design, layout, layering of PCB's, range of functions, etc. Prototype wiremen to a lesser extent, need to share in the range of knowledge of these areas.
2 Microprocessors and digital equipment generally do not fit easily within the constraints involved in audio equipment. Circuits have to be carefully laid out to avoid interference. This complicates the

(lengthening) design process. Only the technical director has adequate knowledge of both audio and microelectronics. He has laid down partition ground rules for the outside sub-contractors. As more PCB work gets carried out in-house, this knowledge must be passed on to the others. The draughtsman was already learning the principles involved.

In the next chapter, we move on to present a set of conclusions drawn from the case studies presented in the previous three chapters.

Chapter 8

Conclusions to case studies

LARGE SITES

We would have expected that small sites would have been more specialised, larger ones more diffuse in their operations – this did not prove to be the case. Just as we found that small sites tended to carry out a large proportion of the design and production steps themselves, and were attempting to increase this proportion in many cases, so we found that large firms were specialising far more.

In a number of different ways, large sites were increasingly concentrating their expertise on certain aspects of product development and manufacture, at the expense of others. This could be termed the 'no point in re-inventing the wheel' strategy. Case L2, for example, would buy in ever increasing amounts of hardware in increasingly complete form, placing all its emphasis on the software and systems design required. This reflects the increasing complexity of products in the areas under review. Whereas L3 provided another example of the same phenomenon, although as a large batch manufacturer it retained an automated production facility on a much larger scale than L1 where manufacturing, whilst hardly automated at all due to the small size of batches, had declined in significance. Company L6 provided an example of a firm where production and design had been placed many hundreds of miles apart, a separation facilitated by the very large batch sizes involved. Designs were provided for the mass market, with only minor modifications, if any, being allowed at the production end, whilst L7 had not seen this shift. Instead, whilst production was maintained on the site, it had become both highly automated and increasingly professionalised. Just as with sites such as L1, almost half the employees at L7, including many in production, were graduates.

Elsewhere, the high cost of both skills and capital equipment were causing companies to use a variety of different approaches, notably inter-company products (L3, L4), international projects (L2), and the sub-contracting not only of production but also of design authority where certain areas of the product were concerned (L5). In this last case, it was the electronics aspects of the mechanical product that were dealt with in this fashion. As microelectronics-based circuitry gradually replaced electro-mechanical controls in the product, and increased the complexity of the systems as a whole, the question of whether the company should not move into the electronics area remained on the agenda. Already micro-electronics-based components accounted for 25 per cent of the whole, and were going to grow in importance. To ensure compatibility with the total product, a high level of understanding of that product was required and those involved in design assurance were expected to have a high degree of hybrid systems expertise. People with this kind of knowledge are rare in the labour-market and are more easily retained if they are involved in producing their own designs rather than assuring someone else's.

Unlike small firms, large sites show fairly rigid stratification between different layers, functions and skill levels. This is borne out in their manpower policies. The higher the level of technology encountered the greater the emphasis there was on graduate recruitment, and the shifts of strategy described above: concentration on design, separation of design and production, automation and 'professionalisation' of production, all meant an increase in the proportion of graduates employed. This point may be legitimate, but companies may be permitting this shift to colour their attitudes so that 'lower-level' skills are seen as being of relatively little importance. The sites where alternatives were being sought for this trend, or where the changes were treated with more perspective, will be described at some length in the chapter on selected training strategies.

The large firms were more likely to be unionised, on the whole. Those that were not were the American multi-national (Companies L6 and L7) and the single-site paternalist firms, Company L3. The effects on the latter of the hiring of previously-unionised labour (largely from Case L2) have been described. In some respects, these changes were part of a cultural shift whereby paternalist norms and practices were being undermined. Changes in technology and product orientation (less emphasis on traditional products and more on those involving electronics) had led to large-scale recruitment of more highly skilled and more formally qualified people. This change in turn meant a decline in the influence of older sections of the firm and less emphasis on seniority. Some of these effects may also have been due to increased size, as Case

L3 had recently made the transition into what we have defined as the 'large' category. Unionisation played a large part in the character of the very large sites, Cases L2 and L5, and large site groups Aviation Ltd and Rangefinders Ltd. In all these cases, negotiations were complicated by the multiplicity of unions (and regional branches of the same union) which followed demarcation lines. These divisions have, of course, become 'part of the terrain' for the parties concerned, with both advantages and disadvantages for all sides. Management strategists could, for example, play on union divisions when introducing new technology, although no clear evidence for this was found. Certainly in the long run such divisions impeded strategies for internal training and up-skilling on the lines proposed at Aviation Ltd, with shop-floor unions playing out the same defensive role as middle levels of management appeared prone to do in a number of the companies studied.

Large sites, by their very nature, have more elaborate administrative politics than small firms, which may impede the kinds of changes under discussion. The broader survival margins of the larger concerns permits them to persist in more wasteful HRM practices regarding human as well as material resources, as the case of Alarms Ltd showed. Large companies in particular create problems for themselves by their very strength in the labour market. The ability of the 'big names' to hire in large numbers of highly qualified new staff means they will be tempted to let skills already in their possession, at the graduate level in particular, go to waste. This was the problem that the Training Department of Rangefinders Ltd had set out to solve. However, the insulation (temporary as it may be) of a large firm to its problems in these areas can be an advantage, since it allows for enough time to take action once the crisis point has been recognised, as appears to have been the case at Alarms Ltd.

MEDIUM-SIZED SITES

Medium-sized sites, perhaps inevitably, present a combination of the characteristics of the other two categories regarding business strategies, HRM, and manpower policies.

Medium-sized sites in the sample, particularly if they were self-contained firms (Cases M3, M4, M5) saw a tension between maintaining their direct control over as much of the process as possible and the concentration on design and development that product microelectronics appears to necessitate. Where a high degree of customisation is implicit in the product market concerned, as with Case M4, this 'broad spread' of resources, assisted by new process technology, appears to present no impediment to success. The case of M3 was similar, although there the rapidly growing electronics design and development section seemed to be leaving the rest of the predominantly mechanical and blue-collar company behind, carrying out design work for outside contracts. Electronics design had been hived off from Company M7, while the original company sought, through new process technology and large-scale training, to reduce the damage caused by the collapse of its traditional market. Similar problems, though less severe, were affecting Cases M5 and M6. In the former, new technology had been invested in, but training initiatives appeared to take place more at the behest of workers who wished to avoid displacement through this same technology. In Case M6 a limited, but highly organised initiative was being tested out with a view to a larger expansion of training activities. As a successful part of a small multi-national, Case M1 seemed to enjoy greater stability, although the 'fickleness' of software engineers appeared to leave them with problems, not least in that it threatened to destabilise the attitudes of other and traditionally more loyal engineering staff.

Case M2 was the only one of the sample to have largely ceased manufacturing, with a large concentration on design and development, production having been reduced to the final assembly and test of bought-in sub-assemblies.

Regarding skill provision, several medium-sized firms/sites behaved like large firms with apparent success, despite not being able to offer the same career progressions. As with Case S6, this was largely due to being involved with the kind of technology that meant interesting work, to high prestige on account of this, and to strong connections with the academic world, although the latter alone could no longer ensure their profitable expansion, leading them to enter more commercial markets. These factors applied to Cases M1, M2 and M4. The sites of Cases M6 and M7 both formed part of much larger concerns and involved sophisticated technologies not applied elsewhere, ensuring that recruitment was not amongst their main problems. None the less both were using training as a means of enticing in recruits who might otherwise have gone to larger rivals. Case M5 tended to avoid graduate recruitment, after the manner of the small firms, since graduates were liable to regard them as a 'first stop' before moving on. Instead, they recruited at a lower level of formal qualification, with a view to maintaining a loyal and locally based workforce. Case M3 was in transition. Whilst not as yet prestigious it had managed to attract recruits with higher degrees, although sponsorship of students through less well-known institutions was a more favoured way of recruiting graduates than the 'milk round' method. Case M4 was now adopting the latter, having emerged from a long recovery period. Growing rapidly after having narrowly avoided collapse five years earlier, it was now poised to enter the 'large' category.

Four of the medium-sized sites were non-unionised, and in none of them did unions play a very significant part.

SMALL SITES

Small sites tend to be more flexible regarding the fulfilment of their manpower requirements. This behaviour appeared to be the case regardless of whether they formed part of a larger concern. In most of those covered, key roles – notably in the 'information technology' – area were found to be performed by people who would very likely not have been seen to be sufficiently qualified to fulfil those roles in a larger company or site. It did not seem to have adversely affected their performance.

Small sites were more likely to train their personnel up to carry out specialised skills. This was true not simply on account of the tradition whereby small firms tend to 'stretch' their personnel and develop skills informally, but as a result of the considerable difficulties small firms encountered in recruiting skilled people directly.

Sponsored training by small sites was often used as the basis for attracting unskilled school-leavers into the company. Often these recruits would leave once their training was finished, in order to join larger firms with better career prospects.

This tendency for people to pursue better career prospects in larger companies was balanced by a counter-trend. It was the way in which small firms sometimes offered a wider range of work for the individuals concerned than their level of training might ensure them elsewhere. Also, if the surroundings were congenial, small firms might claim a higher level of loyalty than larger, more anonymous firms. This fact held true to a greater extent where employees were older and had long-standing connections in the locality.

Where these conditions were accompanied by an unusually high technology product range, as was the case with Case S6, then a small company can recruit and keep large numbers of sought-after graduates and postgraduates, recruiting them on a scale that would be difficult for much larger companies producing similar products.

Small firms may be caught in a vicious circle regarding suppliers and customers – one that is related to the supply and demand of skills. In order to survive, small firms need to specialise in some aspect of product development and manufacture. At the same time, on account of their size they are in a weaker position regarding sub-contractors than large firms are.

They are thus forced, in some cases, to concentrate on what they do best and at the same time cover themselves for as much of the remainder of the work as possible, which can lead to a contradictory strategy. That is to say that, for example, a firm like Case S6, whilst placing great emphasis on design and development, was also recruiting large numbers of blue collar workers so as to reduce dependence on sub-contractors. Case S4 was also attempting to reduce this dependence, which threatened to harm its competitive position on lead-times and delivery deadlines, the pressure of which was also forcing it to take on people without being too selective regarding skills. Case S5 made it a practice to rely as little on outsiders as possible, a strategy facilitated by the relatively restricted role played by microelectronics in its products. Case S3 was a largely blue-collar company which was moving into products involving microelectronics. For the early stages of this transition, it was dependent on research and development carried out in another plant of the same group. It was likely to experience difficulties in recruiting the design and development staff it wanted on account of the limited role microelectronics was to play, and the consequently limited career prospects it could offer in connection with electronic-related skills. Case S1 was unusual in that it appeared to be making a successful, if small-scale, transition from a company that traded on its mechanical design and assembly, to one where the emphasis was on systems design, even though electronic modules were bought in intact. Assembly (always customised) would continue to be crucial, however. Case S2 presented the unfortunate case of a small elite company squeezed between suppliers and customers. In this case, one large supplier provided all the systems hardware, with the smaller company only seeking to rectify this imbalance by buying different systems too late.

As a result the smaller company could not rely on prompt delivery of the essential hardware. At the same time, customers, buoyed up by sharp competition, were in a position to demand too wide a range of services-hardware, software, installation and training. Banks proved unwilling to underwrite the cash-flow complications that resulted.

SUMMARY

The main findings may be summed up as follows: product customisation was found to take place on a large scale on most sites, entailing large requirements regarding graduate engineers (electronics and software) and people with 'hybrid' or systems-expertise; the latter reflected the course taken by product development in many cases. Specialist graduates were unable to cope with the type of convergence between disciplines that appeared to be occurring. It was therefore part of a larger problem – the difficulties encountered concerning the 'socialisation' of graduates into an industrial or commercial environment. Many graduates were too research-oriented or specialised to fit with company priorities. In view of some of these factors, and a bias we perceived concerning graduates (on the part of graduate managers), we suggest a simplified role for technicians, whether this takes the form of devolving functions to the technicians' level, or, perhaps more appropriately in many cases, increased sponsorship of technicians and others through university courses. In addition, we were convinced, through a number of our cases, that one of the solutions to the skill shortage lay in reducing the wastage of skills in companies, whether at the graduate level or elsewhere through comprehensive and regular retraining programmes. Our evidence suggests that the priorities of the recruits themselves will make recruitment more difficult for those companies which do not do this.

Finally, we observed how each trend regarding products, technologies and skills, appeared to provoke counter-trends, whether at the behest of management or otherwise. It was suggested more than once that the strong position of information technology was developed with the aim of reducing the need for them. A related conclusion was the way in which, through the application of design automation and CAD/CAM, companies were attempting to organise small-batch production in such a way that it resembled mass-production.

Thus, optimism to the effect that we are entering a new era where small-batch production and customisation will lead to more skill intensive work organisation, whilst not being wholly misplaced, needs to be tempered regarding the range of possible HRM scenarios.

Chapter 9

Selected training strategies

This chapter sets out to show how the training management function, largely overlooked in the past, can operate as a prime mover in organisational change if those in control of it have a coherent HRM strategy. The experience of the training managers, as given here, also delineates some of the main problems regarding skills in British companies and the wider repercussions of these problems. The companies in this section are additional to those in the main study.

X1: 'AVIATION LTD'

Aviation Ltd, was a large site belonging to a larger concern. Its operations covered research, design and development, manufacturing and testing – all regarding aircraft. It had a workforce well in excess of 10,000, organised by twenty-seven trade unions and staff associations and a financial turnover in the region of £350m in the mid-1980s.

Manufacturing on the site was in flexible small batches and involved the most advanced process technology. CAD, Direct Numerical Control (DNC) machine-tools and FMS were present on a massive scale.

The overall skills strategy

The Training Department was one of three personnel departments reporting to the personnel director, the others being Recruitment and Industrial Relations. Industrial Relations was, it appeared, being downgraded, in so far as its existence as an independent function was concerned. There was also a likelihood that recruitment would be subsumed under an upgraded Training Department.

Training on the site has a full-time staff of eighty-four and an annual budget of £7m, representing almost 2 per cent of turnover and 4 per cent of the annual salary bill.

The problems the department, under new management, were setting out to solve were immense both in terms of scale and in terms of organisational politics. Briefly put, there were too many of the wrong type of skills at the bottom of the skills-pyramid and not enough of the right ones at the top. The training manager's chosen strategy was to use training to up-skill and upgrade the whole organisation, a process that would appear like a cascade effect in reverse.

The process could not easily be launched at the shop-floor level, however. There, management had implemented numerical control machine-tools on the traditional British model (see Sorge *et al*, 1983). That is to say that machines were pre-programmed by remotely assisted production control staff, technicians and engineers. This left the operator in something of a machine-minder's role, utilising very little of the craft skill which had previously been an essential part of the job. Unwilling, or unable, to oppose the introduction of the new technology (including FMS) the shop-floor representatives (largely AEU) at each level insisted that there should be no change in the qualifications for the job – that is, that the apprenticeship 'green ticket' would remain obligatory.

Such a stand is not unreasonable in that it is general trade union policy that new technology should not be implemented with a view to downgrading jobs. It may appear impractical, however, if, as occurred in this case, the policy cannot be altered in response to management moves to upgrade the skilled workers involved via retraining, thus drawing a distinction between the downgrading of the job and the downgrading of the operators as individuals. The problem here is that the operators might then have moved into the more skilled preserve of another union (such as MSF). The AEU (or whichever craft union is involved) would then be left with fewer (management claimed the machine shop was over manned as well as over-skilled) and less-skilled members in the area it traditionally organised. It was not certain that management wished to de-skill the shop-floor so much as move the inappropriately skilled workers out and introduce new mixes of skill. Very likely, the unions were wary of what management meant by the 'integrated' skills that would take the place of the traditional craft skills that currently existed. Possibly the shop-floor would see a polarisation between machine-minders and trouble-shooters.

According to the training manager, the success of the whole strategy for the site depended on moving the craft operators out. Until the shop-floor was reorganised, the benefits of the integrated technologies would not be adequately realised, since they were not being used flexibly enough. The idea was to retrain the craftsmen to carry out machine servicing or white-collar activities such as part-programming, jig-draughting and other technician tasks. From this point, it was hoped that the ablest would then be trained up to fill the specialised posts in the development engineering areas where skill shortages were a serious problem.

This policy appeared to be both logical and progressive to the extent that it formed part of a strategy to become less graduate-oriented in selection. It is, however, interesting to see how the strategy (which may be compared with that of 'Rangefinders' in similar circumstances) takes the British form of skill strategy to unprecedented lengths. Whereas German firms (as reported by Sorge *et al.* 1983), used the craft skills in the company as the foundation around which numerical-controlled machinery would be introduced, British firms' application of similar technology flew in the face of such skills, managements preferring a service-style organisation with skill and discretion in the hands of proliferating white-collar departments. With Aviation Ltd, this strategy was poised to achieve a symbolic perfection – the very people whose craft skills were displaced by the use of DNC in a service-style fashion were themselves to be trained into more white-collar specialists. This development should not necessarily be decried; short of reintroducing DNC machine tools in a craft-oriented way and then having to displace the technicians who currently controlled the process, it appeared to be the most appropriate strategy.

Faced with deadlock at the lower end of the scale, training management looked to ways of starting the process at the top. Rather than a wholesale move of large numbers of people through the various departments they would fill the gaps they had in the more research skills in development by giving special training to people in the technical departments below (which were considered to be overstaffed) and then selecting candidates for promotion on the basis of their performance in training. The disadvantage of this approach is that at each stage the best people would move to the higher-skilled and sparsely peopled departments, something which the managers of the technical departments they had left were likely to resent, leading them perhaps to oppose the scheme. The advantage of the scheme was that some of the skill-mixes in demand cannot easily be met with by universities as they are highly specific hybrids. At the same time, the company had close links with three major universities who ran courses which reflected the industry's needs.

The scheme could also assist with the larger and more serious problem presented by shortages of skills that were not industry-specific at all, but which had companies right across the spectrum competing for them – skills to do with electronics hardware and software. It was thus not so much skills of hybrid content which caused difficulty, but skills of hybrid application, which could be used in a whole variety of industries.

Whilst the mass up-skilling scheme described above was the long-term aim of the training department, its sub-strategy regarding the relation of training to recruitment was its current forcing area. It was an issue which the training manager had managed to take to board level, and had canvassed and received support there, provisionally at least. In the longer term, opposition to both the larger strategy and its sub-strategies was likely to come more from the middle echelons, where managers were attached to short-term criteria in decisions, a result of close budgetary control.

Training and development

The training manager was attempting to reverse two specific trends in recruitment policy, both of which had led to serious failure regarding the fulfilment of company skill requirements. The first trend was towards direct recruitment from universities without the middle phase of sponsorship or company-based training. In recent years, senior financial management had actually cut the training budget assigned to sponsorships, since in their terms it was a waste of money when the machinery for direct recruitment was available. Direct recruitment took place without the involvement of the training function and was carried out primarily through the departments for which graduates were required. These departments were usually managed by graduates who, it was alleged, insisted on employing graduates for all jobs in the department, whether or not such a person was really required.

Both these trends, which had become reinforced in recent years, aggravated the skill shortages – on a national level one might maintain that they were a major contributing factor – not only because the company was having to invest more to recruit people against intensifying competition, but also because the two approaches both led to unacceptably high rates of dissatisfaction and turnover among recruits.

The company was recruiting 150 graduates per year via the direct route. Of these, 120 would go to the Engineering Directorate where they would be employed in design and development, increasingly with an electronics/software bias. The remainder would be assigned to production, quality, testing, etc. For those 150 places, they would receive 4,000 applications. Many of these would be unsuitable; others would be routine applications from people who, spoiled for choice, would often not take up the offer. At least 30 per cent more offers had to be made to secure the right number of places.

For their induction, graduates would spend eighteen months to two years on departmental training/familiarisation, which again would not involve the Training Department. Such training would be geared to the achievement of Institution membership. As with L4, this training served largely as an added incentive to bring recruits in. Unlike L4, Aviation Ltd, was wide ranging enough in its activities, and influential enough with the Institutions themselves, to ensure a useful balance between Institutional and company-specific requirements. None the less, one way or the other this represented a great deal of wasted expense when, as was very often the case, graduates would leave on completion of this training. Reasons for leaving were usually to do with salary and location. The training manager considered it unwise to recruit graduates unless they originated from the North of England (where the company was situated) or had studied there, preferably both. People who came from or studied in the South usually returned there at the first opportunity, so it seemed.

The training manager had submitted an analysis to finance management concerning the costs and benefits of direct recruitment and recruitment through sponsorship or training. If all the costs of actual recruitment and those of induction training were added together so that the cost of labour turnover could be calculated, it could be demonstrated that the latter methods of meeting skill requirements, which entailed a much smaller turnover, were far more efficient. The more cost-effective methods proposed by the training manager were also those which took place under the control of the Training Department. Basically, they involved sponsorship of students, including ex-apprentices, with an emphasis on polytechnics rather than universities, conversion training for non-scientists (this was one of the less successful strategies since the people concerned were often more keen on working in small companies and software houses than in large industrial companies), and the recruitment of more people through sandwich-placements – who would be better assessed by the company and less likely to experience culture shock on their return.

A further method was that of training up technicians to do graduate jobs. One of the reasons why some departments had a high turnover of graduates was, as already inferred, that the recruits were over-qualified for and frustrated in doing what were basically technician-level jobs. Technicians were seen as more adaptable – depending on what training they received they could carry out greater or lesser jobs without apparently feeling the need to see each assignment as a step either towards or away from 'bigger and better things'. If such outlets were opened to technicians it would, in addition, open the way for the larger strategy of mass up-skilling to those who at present were defending obsolete or overmanned positions in fear of displacement. Reducing the number of graduates being recruited did not displace anyone – graduates were in short supply and seemed only too pleased to go elsewhere. Perhaps the most important and most indicative component of the strategy was its proposed amplification of the sponsorship programme, which had in recent years suffered cut-backs. Rather than use direct graduate recruitment as an (albeit indirect) instrument for reducing dependence on lower-grade skills, and hence apparently doing away with the need for an apprenticeship programme, in this case the training manager was endeavouring to integrate graduate recruitment into an increasingly high-skill apprenticeship system.

The company took in 120 apprentices a year, compared with up to 180 six years before. Demand for places was such that 180 applications had to be reviewed, involving a written entrance exam and between 400 and 500 interviews. After one year in which all studied at the same level, apprenticeship would be streamed into technical and craft, with the balance moving more and more towards the former. In addition, 15 per cent on average would be selected for degree sponsorship. Another forty recruits of A-level standard would be brought in each year to enter the programme and be sponsored to graduation. This additional group brought the total number of apprentices at any one time to 700.

The fact that the wages of all apprentices and sponsored students were paid out of the training budget represented a major advantage in the training manager's strategy. Since the departmental managers, to whom he had to 'sell' his scheme would not be footing the bill, he was likely to be given a chance to put his statements concerning the cost-effectiveness of this approach to the test. Such was his confidence in the scheme that he was countenancing financial risks by bringing in more than the quota of sponsored students and farming them out to departments in the hope that the results would more than balance the account. As a whole, this training strategy is noteworthy for the way in which it promises to find a route through the bottle-necks thrown up by the national obsession with recruiting (though not supplying) university graduates and the more wasteful aspects of the service approach to manufacturing. Rather than taking the impracticable step (in the institutional circumstances) of going back to craft-centred design and production, the strategy attempts to make the service approach more effective through reintegrating the craft, technical and engineering skill hierarchies.

X2: 'ALARMS LTD'

Alarms Ltd consisted of a number of sites spread around the country, encompassing a total of 3,000 employees. As a whole, it formed one of the profit centres of a very large British owned concern. The product range of the company included a very wide range of monitoring devices, from simple burglar alarms to elaborate building monitoring systems.

Changes in the technical sophistication of the product-range underlie serious problems the company have been having regarding skill profiles, industrial relations and general morale. These were perceived to be serious, but no action was taken for almost a decade, a situation which manifested itself through a high turnover of board members.

Nine months before our visit to the company's headquarters, one of the company's divisional general managers was appointed as group training manager. The former general manager had previously been a mechanical engineer, and was now nearing retirement age. He was thus seen to have the right kind of technical qualification, to be of sufficient status to increase the standing and effectiveness of the training function, and to be near enough to the close of his career not to be seen as a threat. The description given below, regarding the diagnosis of the company's problems and the means by which they were being dealt with, relies heavily on a presentation given by the training manager (which had been given at all the sites of the company) as well as on discussions with him.

The Problem

According to the training manager, the root of the company's problems lay in the extent to which product development strategies had progressively outstripped the abilities of people in many areas of the business, including sales, design, installation, and after sales service. The outward manifestations of this trend were on the one hand the turnover of the company's directors and, on the other, the breakdown of trust and confidence between workers and management. Management, we were told, were perceived as not being able to comprehend the real needs of the company, its products or its workforce. To the extent that this was true, management was indeed unable to carry out its functions adequately, leading to increasing distrust on the part of workers concerning their job security. The company's relatively safe market position cushioned it from the full effects of this process. Many similar but less well-protected companies have presumably disappeared on account of such sequences of events.

Although the problem had only reached crisis proportions in the early 1980s, when, we can assume, competition began to threaten its survival, it had actually passed a critical point in the mid-1970s. This point was conceptualised by the training manager as being that where an increasing technical complexity first overtook the technical and managerial competence of the company. It was demonstrated by the illustration in Figure 9.1.

The significance of the date 1951 in Figure 9.1 was that it was the year in which Shockley invented the transistor. It will be noticed that 1975 more or less coincides with the appearance of the microprocessor in the industry. Interestingly, technical complexity, building incrementally on the transistor and later innovations, was seen to have been progressively overtaking organisational competence well before the arrival of the microprocessor. One may take issue with this model in so far as relative competence is seen as standing well above technical

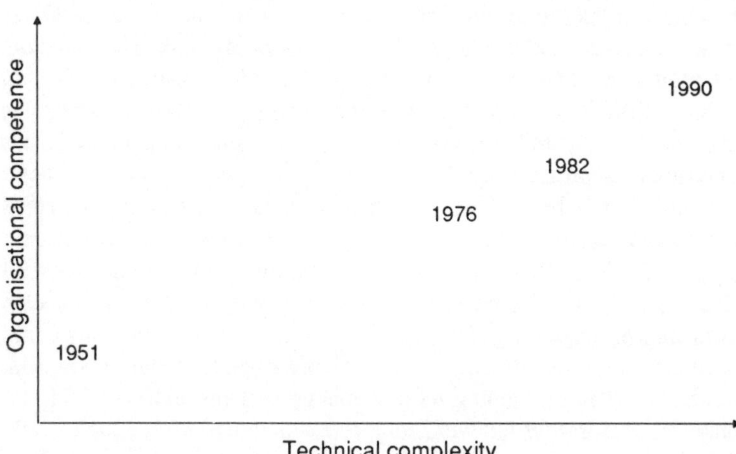

Figure 9.1 Trajectories of technical complexity and organisational competence in the electrical engineering industry (as viewed by the training manager of Alarms Ltd, Mr F. Jane.)

complexity in the decades before 1975. It might have been more appropriate to have seen relative competence rising before falling, although this reflects the complexity of the notion of relative competence in itself.

Where the model seems particularly apposite is in the inference that relative technical complexity may actually feed off the decline in relative competence; that is, the lower the relative competence of the organisation falls, the more technology will be seen as the organisation's salvation and its development will be accelerated further to compensate. In this way, technical development allows the organisation to live on 'borrowed time' (as this organisation apparently had been doing) whilst in fact deepening the problem. It was significant that the training manager was of the view that too much design input in products and processes had been expended on designing skills out rather than capitalising on them. This progressively encouraged learned incapacity on the part of the workforce when there became a problem when the next stage of technical development required a more sophisticated skill input. This in turn would lead to more technical innovation aimed at eliminating the need for skills, and so on, untrammelled by the rest of the organisation, literally spewing out products which the other functions were at a loss to understand.

Although these more sophisticated products were in line with customer requirements, and opened new markets (building monitoring systems as opposed to intruder alarms) they were alien to the products which managers and workers alike had 'grown up' with. The company attempted to compensate for these shortcomings with increases in volume. But it was not a long-term strategy, since the market was becoming increasingly heterogeneous. As it became so, individual sales of customised projects began to account for a larger part of the business; a combined building control and energy management system (such as the firm had been designing and installing for financial institution) could account for 10 per cent of the cost of a building project. The financial risk per project, and the proportion of revenue represented by such projects, was increasing. The more it did so, the less the company could afford to ignore its growing skills problems. Even for volume products, customers were demanding more elaborate systems, alarms which not only issued warnings but were equipped with transmitters which would automatically initiate corrective action. The specificity of these demands had become such that the company had had to establish a Customer Engineering Division dealing entirely with customised systems.

Customisation and technical advance together may also present problems as they split the organisation's product range in two. For example, at one division of Alarms Ltd, with a turnover of £3.5m, £1.75m was accounted for by installation work. Over 1000 different assignments made up 65 per cent of this figure, whilst three large assignments made up 35 per cent. As technical complexity increased, so the relative importance of large projects was expected to increase. Not only did this have important implications for skills and training, but also for cash-flow management. A supply of small 'bread-and-butter' projects and product lines would have to be maintained if the division was not to run the danger of going out of business. Thus a wholesale shift of the organisation towards elaborate system-building on large projects was not practicable.

Overall then, the training manager was able to conceptualise the progression of the skills problem in the following manner:

1 The rate of technological development overtakes awareness.
2 The scale of the backlog in orders becomes intimidating.
3 The existing educational institutions are overtaken.
4 The formal education of those employed becomes inappropriate and inadequate.
5 The problem is made worse by the fact that the employees have limited time available for training and are unaccustomed to learning new skills.
6 The workforce thus becomes demoralised.
7 The growth of protectionism as a defensive strategy follows.
8 Product design becomes geared to finding ways to compensate for the lack of skills available.
9 Training (such as it exists) is hampered by the lack of educational attainment already achieved, producing a spiral of inadequacy.
10 Overall intellectual levels are undermined. The 'systems comprehension' needed for the new range of products is scarce among the workforce or management.
11 A condition of 'anxious obsolescence' is engendered which finally inhibits any further investment in 'state of the art' techniques.

Technology, products and markets

The question of changing markets needs some further discussion. As long as the products were conventional alarms, they were safely linked into a world-wide market. With the gradual break-up of these mass markets, it became imperative to seek out new markets into which adaptations of the product through flexible technologies could be sold. This entailed considerable preparation; not only did the new areas have to be identified in advance, but information of a new type had to be collected on these areas so that products could be designed and manufactured to fill their needs. Poor co-ordination at this stage led to confusion of whether the company was to be primarily market-led (the emphasis being on the selection of new production niches) or engineering that moved ahead of marketing – the latter function failed to understand or capitalise on the advances occurring in the development engineering area. During the late 1970s, a series of arguments took place in the arena provided by the company's regular product review committees. The managers who went on to become training manager and computer-aided engineering manager at that time formed the nucleus of a group which was trying to force company directors to make suitable use of the technical advances occurring in engineering both inside and outside the firms. The particular advances concerned with at that stage were the availability of more power in proportion to the physical size of product components, the advent of integrated circuits and the breaking of the development constraints associated with the thermiotic valves and traditional electronics.

Not only the board of directors, but also the workforce were mistrustful of the immense changes these innovations were seen as entailing. Changes were likely regarding the size, skill, structure, orientation and even geographical location of the company. The training manager claimed his ambition was to see the changes through without the drastic measures he had witnessed at a similar juncture with his former employers, NCR. When electro-machines were superseded in that sector, the company had shifted to California rather than push through what they saw as the necessary changes in an environment constrained by a high level of unionisation. Many managements it seemed, associated newer or more flexible technologies with a de-unionised organisation. A similar example had been the move of a typewriter company to Spain in order to make electronic typewriters, thus closing factories in two towns in Britain. The result of this process, it was said, was that a spiral of bad industrial relations was set in motion. Unions and their members would suspect that, with any wide-reaching change that was imminent, that management would be preparing to 'rat' on them. In turn, they would become doubly uncooperative with a company which they no longer perceived that they had any common interest with. They would also become more closely identified with existing technologies and practices as their only means of defence. In the long run, this attitude made conflict and marginalisation even more likely for the workforce.

The training manager's view was that it was highly irresponsible for a company to desert its traditional base in order to change direction, and that such changes could be adequately implemented with the existing workforce if common interests could be recognised all round, and morale increased. Training was seen as the most suitable direction through which to attack this latter problem.

The crisis, as far as the training manager was concerned, arose out of the boom in manufacturing technology which had contributed to a halving of the manufacturing workforce nation-wide. The inability of those in authority to cope with these developments has led to a collapse in the trust placed in political and industrial leaders. The crisis has been deepened by the influx of foreign goods produced by cheap labour, and by more advanced technologies and forms of work organisation. With this breakdown of trust becoming increasingly severe in the late 1970s, the government's policy was seen to have artificially worsened the effects of the recession in order to secure through fear the compliance and mistrust which was lacking. It could also be said that the recession provided the screen for the moving of many manufacturing operations abroad, very much as in the NCR example from the 1960s. This policy was not seen as providing a viable long-term solution to the problem. In the alarms industry, the legacy of fear did not give a strong basis for co-operation, but merely passive resistance, paralysis and calls for protectionism.

The response of management had thus been to make the active co-operation of the workforce as minor a contingency as possible. Production line work had been de-skilled and simplified through a series of deliberate design initiatives. 'Idiot servicing' had been introduced to reduce the dependency of management on diagnostic skills. The current vogue for training was, then, a belated recognition that high-technology industries could not in the long run be effectively staffed by idiots, and represented a tentative reversal of the reductionist thinking that had dominated industrial policies for decades, culminating in the early 1980s.

The elimination of labour and the active reduction of skills requirements had produced its own spiralling effect, which went as follows:

1 Employees start out from a background characterised by poor education generally.

2 Their training in companies is often limited to training for the production or operation of specific new products and processes, a form of training which neglects general principles which can be used as a grounding for further knowledge.
3 Due to their poor initial education, employees are unable to comprehend the principles underlying their brief product-specific training.
4 They become disillusioned with the company, and are convinced that they will not be able to adapt to change. Change will thus be perceived in negative terms.

Such a spiral was said to affect both workers and management, and to tie them into the condition referred to as 'anxious obsolescence'. In a sense, the training manager's aim was to attempt to smooth over the cracks produced by the change process through the application of comprehensive training. We should say that fear of change is not primarily a psychological state – we would say rather that it is the natural result of having vested interests in a particular *status quo*, whether regarding product technology, work organisation or even company culture. However adaptable someone who has spent years or even decades might be, the chances are that someone abler or merely younger and less steeped in the existing system may be in a position to profit more from the change. In the short term, it will be in the individual's interest to obstruct such change on these grounds. In the long run, the collective interest cannot be so served. Training may thus be provided as a kind of 'organisational lubricant' to enable the more easy adjustment of power bases so that, even if such bases will be in many cases reduced, the long-term losses for those concerned will be cut. It also provides a means of more systematic selection for those who may hold greater bargaining power in the changed *status quo*. Training may thus be used to give such a shifting of the skill/power structure in a company the legitimacy of an egalitarian approach.

As a final note on the seriousness of the problem, we may recapitulate with the following observations:

HRM strategies which had been predicated on the need for minimum skill-levels produce generations of employees and managers unable to grasp the principles behind the processes they are dealing with. When advances in technology lead to the integration of part of the whole of such processes, as has increasingly been the case, they are at a loss, and company performance suffers, along with the individual's or group's security of employment. People who have 'graduated' through the impoverished education spiral described above cannot comprehend the complexity of the interfaces between the vast number of control mechanisms involved in the newer products. This effect leads to damage-limitation strategies and piecemeal technological solutions which have little long-term value. At the same time, the transmission of signals to and from remote end terminals presents a problem for fault diagnosis. Technical faults are not often such that they have their origin in places discrete from where they manifest themselves. In order to ascertain the origin of such faults, far higher levels of diagnostic skills are required than have become the norm in British industry. In the training manager's phrase, 'people who have for so long been treated as idiots become idiots'.

Attempts to remedy this situation, particularly against the backdrop of the 'information society cult', tend to concentrate, as we might argue in the vein of Sorge *et al.* (1983), on the assembly and communication of pure facts. This approach takes little account of the fact that many of those for whom such information is intended have no adequate frame of reference in which to place these accumulated facts. It may be even more true of some groups within management than it is for maintenance of other shop-floor groups.

The result was, then, that 90 per cent of the company's employees, from the board of directors down, were, in what the training manager considered to be a conservative estimate, incapable of understanding the products the company were producing. When we consider how the latter were becoming increasingly complex in line with their being tailored to individual customer requirements, the position may appear even worse.

The training initiative

The training manager had apparently been in the process, soon after his appointment, of developing electronics teaching modules which would make available to those whose technical background was lacking or obsolete, the basic principles of the systems now being dealt with. He then discovered that the then Manpower Services Commission 'pump-priming' initiative (Open Tech) had led to a similar project being developed at a polytechnic in the region. This course included 140 modules for self-tuition.

Two major obstacles stood in the way of such a scheme being implemented on a large scale in the company. On the one hand, people refused to admit their ignorance, as the training manager termed it. On the other, they refused to submit to the kind of assessment whereby their ignorance, or need for training, might be ascertained. For the first, the training manager made his being allowed to purchase the appropriate modules a condition of his staying on in the job (here his near-retirement age gave him an advantage). For the second, the modular structure of the training, by which one could only move to a higher level after passing a lower level, and so on, meant that assessment could in practice be an automatic by-product of the training itself. It could, of course, only be so if it was taken up on a sufficient scale. This was partly helped by the fact that one of the major unions involved, the EETPU, or rather its staff section the Electrical Supervisory Staffs Association (ESSA), was also making available modules of a similar type for 'open learning'. The training manager offered to have the two courses implemented in parallel, which the union accepted.

In the training manager's opinion, the ESSA course was ponderous and of a much less useful type than the Open Tech course. This would not have surprised our MSF contacts, in whose opinion the (large-scale) training activities of the EETPU and its subsidiaries represented a fundamentally political initiative to enable EETPU members to move through their own union's training system into the skill preserves which might have been seen as belonging to other unions, such as MSF. At the least, of course, the EETPU could be credited with having understood early on, the new potential for training to be used as a wide-ranging strategic tool. In so far as the ESSA course was aimed at the upgrading of its staff rather than increased effectiveness of training, this could be said to be borne out by the fact that it was highly technical in its language and its format, whereas the plainness and directness of the Open Tech course presentation was what recommended it to the training manager, who was afraid that potential students would be alienated by the wrong register of language. Whilst the ESSA course emphasised the learning of facts, the Open Tech course encouraged the application of rational thinking to simple electronics problems, which was seen to be the main shortcoming of the employees in practice. The training manager made much of the fact that in his view the majority of employees were not intrinsically less intelligent that their organisation superiors, and that developing their capacity for 'rational thinking' would have considerable benefits.

The course was so designed that the successful completion of each module would be met with a certificate from a local technical college. This was seen to ensure that competitiveness amongst employees would overcome the scepticism many had for the scheme at first, and give it a momentum of its own. It was not only competition that would have this effect, but the realisation that their jobs might depend on their keeping up with others doing the course. The course was not, however, open to everybody at the same time. Groups of workers were 'targeted' for the course, and as each group gradually became involved, the information concerning their progress would be compiled to form something resembling a directory of skills, something which had not previously existed in the company in any form. The course would ensure that this directory was compiled on the basis of homogeneous standards.

The training manager believed that the training programme, being an active initiative rather than some passive exercise of damage limitation, would provide an outlet for the discontent that had been seen as a major problem in the company. Already, he could say that there was evidence that those who had been most vocal in expressing their discontent on the shop-floor were the ones who were taking most enthusiastically to the scheme.

He took particular satisfaction from a recent conference of all the unions involved with the parent company. At this event, electricians' union shop-stewards from Alarms Ltd, who had previously been hostile, were heard to be praising the training scheme. This was seen as proof of how the right training scheme could heal apparently intractable problems. It was stressed that the scheme was not an attempt to circumvent union organisation, but had been formally agreed with all seven unions involved.

The scheme was intended to cover 100 employees within five years – that is, one third of the total workforce. So far it had been targeted and taken up by eighty-five people, located in the sales, installation and servicing areas. Several hundred more were expected to be involved within two years, by which time the more able of the first batch would have completed all the thirteen modules. The figure of 1,000 represented all those who came into contact with electronics in the course of their jobs.

The cost of the project for the first year was £47,000 (the modules cost £150 each). This was seen as minimal compared with 'the cost of ignorance' which the company had been, and was still paying. The ESSA scheme, which involved thirty-five to forty people being sent on courses at one time, cost £300 per person and took the employees concerned away from the job for two weeks. The Open Tech scheme was seen as representing sixteen hours' work per module on average, all of which would come out of the employee's spare time, not the firm's. The Open Tech scheme was clearly the more cost-effective from the company's point of view. The training manager's enthusiasm for Open

Tech drew on his view that training employees should be the responsibility of the company, not of educational institutions or trade unions. Though this should logically mean time being allowed off for study in company time, this would increase the cost of the scheme and give senior management more arguments to use against it.

Training had an overall budget of around £0.5m in the firm. The other part of the training strategy was to use the remaining 90 per cent of the budget more effectively. The training manager had calculated (no one had seen fit to do so before) that the existing network of four training schools which ran the product training courses and budgeted at 100 × 2 man-weeks a year, represented 4.3 per cent of the total payroll each year. Sixty-three man-weeks were required to train each instructor who was only given thirty weeks' work a year. The problem of having so many far-flung sites meant that each region was unable to fill a classroom for any one course. It was thus seen that resources already given to training were not being used to capacity. Changes would have to be made so as to justify financial assistance for the Open Tech initiative.

The board of directors were seen as having little understanding of the need for training. Since they always preferred factual evidence for decision-making where finance was involved, the training manager was in the process of attempting to compile a figure for the rectification costs which could be associated with the lack of training in the past, a sum which would dwarf the existing training budget and help justify further investment. As far as management was concerned, the aim of the presentations around the country was to convince them of how skill-wastage was central to the company's current malaise.

Interestingly, it was a social/educational divide rather than lack of technical education that was seen as holding back the board's understanding of these matters. All had highly advanced technical qualifications. This led them, we were told, to understand even less the problems technical change held out for those with poorer qualifications.

Also it could be claimed that a high level of specialised technical knowledge did not provide the best basis for understanding the impact of technology on the organisation. This may have lain behind the low rate of growth of the Customer Engineering Department, which numbered around 300 people, well below the level its long-term importance would recommend. The department was concerned with building control equipment out of 'building blocks' in line with specific requirements provided by the customer. The newer markets, such as the building control systems we referred to earlier, can only be penetrated through customisation.

Managers and specialists in this area of the business are usually university graduates, of which the company takes in about fifteen a year. Whilst these are expected to provide their own technical updating once their induction period is over, some provision was being made to cover for the serious shortfall in systems understanding. No systems training had ever been given by the company before. Under the Open Tech scheme, some graduate-oriented modules had been provided. The training manager's preferred strategy however, was to train the field service workers of the Custom Engineering Department in systems understanding via the thirteen modules and in this way 'teach back' into the department itself. The training manager was of the opinion that technical graduates tended to be too specialised in their discipline to be well suited for systems work, as opposed to the design of sub-systems. For the overall systems approach, he believed a good HNC technician with 'jack-of-all-trades' abilities would be better suited.

This view sums up a trend encountered elsewhere, whereby training managers believed that where systems-oriented or 'hybrid' skills were concerned, the best place to develop them was not amongst the technically best qualified, but among the intermediate range – notably technicians.

X3: 'RANGEFINDERS LTD'

Rangefinders Ltd was a major British-owned engineering firm. In this study, we are concerned with one group of sites within the larger company. This group, managed as a distinct entity, was less than forty years old, during which time it had grown considerably. Over that period, its (highly specialised) small-batch product range had seen a significant shift from mechanical and electro-mechanical engineering to electronics.

The training manager at Rangefinders was responsible for all training activity undertaken within the group, covering 7,000 employees (around one-third of the total employed by the parent company as a whole). Prior to this appointment, he had been group personnel manager. This job had involved negotiating membership agreements with thirteen shop-floor unions, and it had been during the period of contraction in the company, coinciding with the 1970s aerospace recession. In the early 1980s, with employment in the group on the increase again, he moved to the Training Department, with the aim of 'reinstating' it after the period of stagnation. The move was seen by many at the time as a demotion, although this perception has been proved to be largely unfounded, given the 'forcing' role training has acquired in recent years (of the preceding two cases).

Training finance

The official training budget was £2.6m, or 1 per cent of group turnover. It covered training overheads charged directly to the Training Department. In fact, the overall expenditure on training is at least twice as high, since, with the exception of the apprenticeship scheme, the cost of training courses is paid by the departments in which those who are to receive the training actually work. All training expenditure passes through the training department in one way or another. Even where line/departmental managers wish to send someone on a course at another institution, the fees have to be authorised by the training manager, who may take the opportunity to place the people concerned on an internal course which the line managers were unacquainted with. Training overheads are provided through a levy of the forty-three profit centres involved in the group, so that, whilst the Training Department is operated on a commercial footing, it is designed to act as a service to line departments and may not sell services outside of the group. It is perceived to be a line management's responsibility to assess skill needs and maintain a skilled workforce, and the role of the training function merely to facilitate this.

The financial arrangements were to work in the Training Department's favour despite its apparent reliance on the perceptions of line management. Since the overheads are paid in any case by the forty-three profit centres, it is in their interest to use them as much as possible as the payment of such overheads means that courses arranged under the auspices of the training function will be significantly cheaper than those arranged outside.

Training appears to have a vested interest in keeping the real total of training expenditure discrete from those who might use the real figure as ammunition and a reason for cutting it back. There is a tendency for individual line managers to bemoan the paying of the relatively insignificant training overhead whenever their profit centre nears unprofitability. Perhaps for this reason, whilst a computerised register gives details of all who have undergone training within the company since the 1970s, no record of the total costs can be obtained. If fees for in-house courses are included the figure was £5.5m, to which may be added the cost of external training courses, for which no centrally compiled figure exists. Informal training on-the-job, which the training manager maintained was the most effective method, was, of course, uncosted. In areas such as the drawing-office, supervisors were continually training people and the results were considered satisfactory. The Training Department thus saw its role of providing formal training, a second-best option, only where the informal system was unable to cope. As we shall see, a major area where such informal training could not provide the results was in the graduate engineering sector, the target of a major training initiative by the training manager.

The training function, which had recently received a substantial grant from the European Social Fund, monitored outside courses to ensure value for money. Students were expected to report back on the quality or usefulness of the courses concerned. One reason therefore why external expenditure had to be approved by the department was that this prevented people being sent off a second time to extravagantly advertised but sub-standard courses run by commercial training companies.

Apart from its commercial-style relationship with line departments, the training function reported to nobody (not Personnel, for example) except for the assistant general manager of the group, who was obliged to approve the training budget. In terms of staff, the training function covered eighteen full-time instructors, 120 part-timers (available from other departments), three supervisors for the apprentices, three administrative staff and the training manager.

Apprenticeships

Rangefinders were (for the sector their business was in), untypically committed to the traditional apprenticeship system. There were 420 apprentices in the group, a number that had not varied (although it has not increased in line with the expansion in the company's employment). Since 1943 when the group was founded, 3,825 apprentices have passed through the system, 1,800 of them still working for the company, many in senior positions. However, although the scheme has in itself continued to flourish, its annual intake now represents little over 1 per cent of the workforce, whereas in previous years it would have been considerably higher.

The apprenticeship lasts four years, of which the first year is spent entirely in the Training Department. Trainees are then assigned to departments and work on a day-release basis. Like most British apprenticeships, the scheme divides after a year into craft (ONC) and technical (HNC) courses. They are also divided into mechanical and electronic streams. The predominant trend is towards electronic/technical apprenticeships.

1	Mechanical 40%	Electronic 60%	Total 100%
2	Craft 30%	Technical 70%	Total 100%

The breakdown above refers to those coming to the end of their first year of training. All apprentices are school leavers. Although four O-level passes are usually needed for admission to a technical apprenticeship, craft trainees may be upgraded. Electronics craft is a declining area due to automation in the area of system-testing. Electronics technicians are needed in increasing numbers in order to have people capable of interpreting the increasingly complex instructions and specifications put out by design engineers. Technicians are also preferred on account of their greater flexibility compared to graduate specialists. The training manager explained his view that whereas university-trained scientists tend to waste away when not engaged in challenging work (so that they were unable to return to their former level of ability when required later), technicians tended to have lower expectations. They could rise to the challenge of demanding work and then return to routine work when it was completed. With increased automation in the form of CAD/CAM many white-collar indirect staff have been removed from the machine shop-floor and redirected to the development labs where they would provide a service for design engineers. It seemed likely that from this beginning a move could be under way to return to the long-discarded form of development organisation, whereby technicians carry out a higher proportion of development work, assisted by design engineers acting as consultants who provide the specialist input where necessary. This trend is likely to be encouraged by the increased use of Computer-Aided Engineering (CAE). With this system, now already on stream, the general manager had instructed group engineering managers that the company's main products need no longer be designed from scratch, but that maximum use should be made of 'repeats'. It would lead to a declining volume of work for highly specialised engineers. Much of the design work could then be divided between able technicians and CAD draughtsmen. The complexity of the design-skills question will be discussed in further detail later in this chapter.

These considerations enforce our view that the training manager had particular ends in view for keeping the apprenticeship scheme intact, even whilst shifting its emphasis. The high number of mechanical trainees may also be somewhat surprising, given that the company's mechanical input had declined in significance in recent decades, and that recruitment of mechanical engineers was relatively low.

What it reflected to some extent was the training manager's view that as software became standardised and application-specific chips became available, the need to recruit large numbers of electronics hardware/software engineers would decrease. The emphasis on design and quality would shift back out to the 'packaging' of the product, just as had already begun to occur with the emergence of hybrid specialisms such as that of 'microprocessor engineers' who were skilled in placing chips in a design but who knew little of the make-up of a microprocessor. This return to the periphery of a product was expected to lead to a renaissance of the mechanical engineers, it seems to us that the training manager intended to have the technicians and craft workers being trained under the apprenticeship scheme available for when such a shift of emphasis occurred. Like many of the senior managers the training manager was a non-graduate mechanical engineer by training. When another mechanically trained senior manager heard him put forward this view, he told the training manager that he considered it a 'little optimistic'.

The apprenticeship training programme embodied certain other views of the training manager. As with 'sandwich' students, of whom around fifteen were brought in each year, apprentices received a maximum amount of hands-on experience in their first year. The training centre was equipped as a medium-sized factory with this in view. Trainees would begin by being trained to use outmoded equipment and methods, and were not allowed to use modern computerised equipment until they had mastered the principles of the old. Like other engineer-managers, the training manager had a 'realistic' approach to computers. It was realised that, particularly with complex small-batch work, it was essential to people who understood the principles behind the computerised process since the latter so frequently made errors. Managers at the company considered 'right first time' to be very much an advertising slogan. They considered from their experience that 'right fifth time' was not on average a bad showing. The consensus was that computer-based skills were 'fragile' in this kind of engineering context, alienating the operator from what the process was actually about. The computer was a tool that was no better than the person using it, and was not to be allowed to structure that person's approach to the job. 'We try to teach the apprentices right from the first that computers are stupid', the training manager declared. This aim lay behind getting them to write their own programmes for CNC machines early on. There was also another purpose for this training, as we shall see later on.

Trainees were also educated not to trust suppliers, this being done through their entering the EITB competition in which they had to design and produce a product in 121 weeks with no official assistance. In the same process, they are taught to work in project teams without any direct supervision. This appeared to be in line with the trend away from traditional line-based structures within the company.

As with L4, attempts had been made to recruit women apprentices, with representations being made to schools etc. It appears that, as elsewhere in the aerospace and defence electronics sector, the company had a male-dominated culture which led potential women recruits to believe they would have few possibilities of promotion. Only six out of 114 trainees were women. Very likely the company did not try as hard to recruit them as they would have done had they been suffering from skill shortages. As it was they were one of the largest and most advanced companies in the region, a region which also contained a large number of universities, technical colleges and other higher education institutions along with a high rate of unemployment. Thus, filling vacancies of whatever type presented relatively few problems.

Importantly, both mechanical and electronic apprentices were made to spend two months of their first year working on each other's discipline. It was considered that, without the right level of integration between the two disciplines in the product it would be impossible to achieve high standards without some measure of 'hybridisation' of skills, at least to the extent of understanding the principles of the other subject. The electronic side of this exchange, the eight-week basic electronics course, was available to other employees outside of the apprenticeship programme.

All apprentices learnt the rudiments of a wide range of skills that they would not be necessarily expected to use in their career: manual draughting, CAD draughting, CNC, assembly and testing, Basic programming and the use of the company's own software. This last was likely to grow in importance as all information, whether engineering, financial or otherwise, was currently being linked into a group central database into which employees in all relevant departments were to have access, both to enter and call up information. The driving justification behind this move was to link CAE information with costings, so that designs could be costed.

Both the use of 'obsolete' skills and the central CAE/costing data-bank were directly relevant to the fact that a large part of the company's business had long consisted in updating equipment they had sold to customers many years previously. The information on all such sales, and its constant modification, meant that they now ran the largest publishing/library concern in the region. Furthermore, almost all their output lay in customised small-batches, either tailored to their own internal requirements (they manufactured much of their own process equipment, owing to the very specific constraints of their product parameters) or those of their customers.

Graduate recruitment and training

The training manager at Rangefinders was a strong critic (and apparently well known as such through his membership of regional training boards), of two widely held views regarding graduate recruitment. On the one hand he, together with other managers interviewed, disapproved of the vogue for 'information technology' degree courses. As we have inferred, the skills involved were seen as 'fragile' when applied to manufacturing, where the essential nature of the process involved constraints which pre-dated information technology, and regarding which computing skills on their own were of little use. The vogue also all too often by-passed the need for systems engineers, the one area in which they had real and sustained difficulty recruiting. Instead, new forms of specialist were produced by the education system. The proliferation of the term 'information technology' in syllabuses was, to some extent, seen as a legitimating exercise. Institutions would use the term in order to secure state funding. The state, in its turn, needed to be seen to be assisting industry by providing skilled recruits.

The second, perhaps more important area of disagreement was over the 'supply side' approach to graduate shortages typified by the Butcher Report (DTI 1985). Such analyses placed far too much emphasis on the production of new graduates whilst ignoring one of the major reasons for the skill shortage – the wastage of talent already existing inside organisations. This criticism referred also to large numbers of non-graduate employees who could be trained up, but in particular it was the specifically gradual dimension of the problem that was the target of the training manager's current initiative.

Increasing the supply of new graduates would in the long run act only as a palliative and, at worst, would add to the existing problem. As technology advanced, graduate 'half-life', the period in which they could give their optimum contribution without retraining, grew shorter and shorter. In large organisations such as Rangefinders, alienation and the sheer difficulty of finding appropriate work for all graduates could render it significantly shorter.

The speed with which graduate potential is exhausted can mean that the combination of effective graduate ability and several years' experience becomes rare in practice. As the training manager maintained, an engineer with fifteen years' experience has something that the most able new graduate lacks, and which cannot be imparted by formal training. In contrast, the 'state of the art' knowledge of the graduate can, in theory, be taught to experienced engineers.

The group employed over 1,000 graduates, of whom the following breakdown can be made:

Electrical/electronic (inc. software)	700
Mechanical	120
Others (Maths, Chemistry, Physics, plus professionals)	180
Total number of employees	1,000

The graduate intake was almost entirely electronics hardware and software. Little distinction is made between the two, with 'firmware' (combined hardware and software) being the company's definition of electronics. Although all the main products are driven by 'embedded' software, very few pure software specialists are employed (a mere twenty to thirty altogether). Software engineers are expected to have a thorough knowledge of the hardware, and how their 'part of the package' fits into an integrated whole with the rest. Despite recent improvements in the provision of firmware graduates, universities were criticised on their leaning towards separate courses for electronics hardware and software, which reflected, so it seemed, an insitutional bias towards specialists: 'it will kill us if we only employ people who can only do one job'.

As we have already described, certain trends were seen as undermining the position of electronics skill in the company. Software and circuit design were increasingly going to become a matter of 'off-the-shelf' buying, even if it occurred within the company itself. Application-specific circuits were soon to be produced by another group within the parent company and would enable circuits to be made specific through the mere passing of a signal through them. This result was seen as going against all the company's traditions of painstaking design work. It could also leave them with the problem of what to do with several hundred electronics specialists. As it was, they were still recruiting them in at the rate of 100 a year, with another 180 being sponsored on sandwich courses. The more pressing problem was how to update those already in the company, and it is to this initiative we now turn.

Graduate retraining

The training manager stated the problem thus:

> we have a thousand graduate engineers and many are already passengers – many will remain passengers for twenty years. If this continues we will be in dire trouble. The skill needs of the company will have to be met through the recruitment of vast numbers of new graduates who will not, and cannot be expected to, exist in sufficient numbers. To interpret that as a shortage of new graduates is to miss the point entirely.

Only a small minority of engineers could be expected to keep up to date through private study, whether through open learning or otherwise. It was the company's job to see that the rest were retrained, and minimise the skills haemorrhage. We could point out that the beginning of the problem as in a number of the companies studied, lies in a mismatch between the individual's and the company's expectations. The personality types best suited to success in a university environment do not always adapt successfully to life in a large manufacturing organisation. It was, of course, emphasised that many did adapt very successfully, and that those who achieve most in one environment often achieved most in others. It remained, however, that many new graduates were, in the company's view 'living on another planet', although, it was said of these, 'but make no mistake, we do need and always will need people like that'. The problem was perhaps not with the more extreme cases, who, as brilliant workers in their specialised field, could always find a niche in a consultant capacity within the company. Less brilliant, but equally specialised engineers were the ones which presented a problem. The source of the difficulty might perhaps lie in the fact that success in university was achieved by individuals to the extent that they resembled the 'brilliant specialist' paradigm. For those who attained the limits of the model uses could always be found – the others, on the other hand, represented the worst of both worlds. Too specialised to have the flexibility of technicians, they were not outstanding enough to fulfil their promise as specialists either.

Skill-needs in engineering are anticipated and planned for in advance through the offices of the Graduate Steering Committee, made up of senior engineers. Areas of expertise are selected which those concerned feel the company will have to move into if the company is not to fall behind. Courses are then devised for these subjects, using consultant specialists in these areas, who are already on the company payroll. This system has been found to be insufficient for the pace of change.

Under the direction of the training manager, therefore, links have been formed with a wide variety of universities and technical colleges found in the area. These links are of a different type to those formed before, in the sense that now the idea is to have lecturers from the universities teach courses on the company-site rather than send the engineers to the universities. Not only is the new system less expensive, but the course is ensured to be more company-oriented than it might otherwise be. A point is made to have at least one presentation on each course given by an employee of the company. Further to this, it is intended that the outside lecturers gain a closer understanding of the needs of the company and the context in which the knowledge is to be applied.

The courses are designed to run one day a week for between six and ten weeks. Subjects covered so far include radar systems, digital techniques and the design of power units. The courses are attended by fifteen engineers at a time, with a total of 300 having attended at least one course over the three years. All those concerned had between five and twenty-five years' experience in the company.

The training manager had long since ceased to use formal appraisal systems as a means for gauging training needs. Apparently in practice supervisors had used a kind of code which had nothing to do with real training needs. For example, the comment 'this man could do with some management training' was taken to mean that the employee in question was anti-social. Two-way appraisals of a sort were organised by the Graduate Steering Committee, whereby those with over ten years in the company would be gathered for a social event where they could discuss their progress.

It should be clear that the company was given to long-term planning regarding personnel and skill requirements. This arose from the nature of their manufacturing business. The latter was largely project-based, and it was possible for an engineer to work solely on one project for several years, after which he might be of little use for any other type of work. In general, the company aimed at a measure of continuity amidst the multiplicity of shifting projects and technical changes which they dealt with. Long cycles of change were preferrred, with no redundancies, the maximum use of old-fashioned skills where appropriate, and retraining or new work for all those displaced. There was thus a wide variety of other training activities taking place through the Training Department, as well as on the job. Apart from apprentices, 600 people will attend at least one course lasting at least one day in the training centre. Many attend one-day courses, including children from local schools who are given an impression of what manufacturing work is like.

Management training is not organised through the training centre, but through the parent company as a whole. Supervisors and project leaders from the entire group are given three-day courses in the centre, 400 in the current year. Graduate induction involves management decision-making games over the first six months of employment. These are intended to teach group co-ordination as well as supervisory skills.

The themes we have introduced and discussed with reference to Rangefinders Ltd's training strategy will, along with those of the other two major cases, be taken up later in this chapter, which will link together the data from all the companies investigated, and of which the overviews were set out in the previous chapter.

For the remainder of this chapter, we shall demonstrate the complexity of the relationship between company training strategy and the forces involved in product innovation process technologies, work organisation, union recognition agreements, and management structures. The account below of the implementation of CAE technology (and, more importantly, its philosophy), at Rangefinders, presents the relationship as being complex enough to undermine conventional notions of 'management strategy' and replace it with a more realistic picture of how differing interests are expressed in an organisational setting, as well as the role that questions related to skills and training play in the process. The account naturally takes up themes outlined in the earlier section on Aviation Ltd, which in turn built on themes broached in the earlier Anglo-German study (Sorge *et al.* 1983), demonstrating how that and the present study may be linked.

Design and production at Rangefinders

The first numerical control (NC) Machines were introduced in 1965.

They have, as the recruitment figures quoted earlier suggest, made considerable inroads into the nature of shop-floor craft skills and the number of craft workers required. We were, for example, shown components which, we were told, would have taken two to three weeks for a skilled worker to produce, whereas now NC machines could do the same in six hours. The training manager did not see this passing of skill as regrettable – he said it would have driven someone mad to do that kind of work. The machine shops are concentrated next to the drawing office and CAD centre. The machine shop includes eight flexible machine centres. The company identify CNC (for which the machines are equipped) as being the emphasis on operator-programming in the utilisation of the system. Given the small batches required the company have chosen the more 'traditional' NC route. With this a pallet is loaded into the machine, sets the switch which identifies the job, loads the tape and sets up the program. The system can be used to bring the raw material to the job, although it has not been up to now. The tape in question comes from the drawing office. The tape can programme which function should be carried out without intervention from the operator, although in practice this is something the operators insist on retaining. The company, on the other hand, are looking for complete automaticity.

This system is, they claim, much quicker and easier and is particularly suitable for repeats. The use of sub-routines makes it possible to run-off an uninterrupted stream of different small batches in a very short space of time by capitalising on what they have in common. NC has enabled some products to be made which would never have been possible before with anything like the accuracy. One example was the thin laminators which are stacked in the radar systems. These are punched with slits that are equidistant and are accurate to plus or minus 0.001 cm. However, such products are required in very small numbers. The size of batches is the main reason for the company's rejection of robotics which are not seen as flexible enough for such purposes.

For some of these components – such as the slotted flat-plate aerials for the radar systems, design becomes a 'mere' mathematical exercise. All that is necessary is to feed the basic specification to the computer. The computer already has stored within it the background specifications for that kind of product. The computer then sends back a print-out which 'suggests' a suitable design which the design engineer then checks. The information is then put on a tape and sent to the flexible machining centre where it is automatically produced. It should be stressed that this is only possible with certain components and certainly not with whole systems.

The company do however seem to be embarked on a course of some conservatism in design, presumably with a view to reducing lead-times. We have already cited how the general manager of the group had impressed on his senior management that nobody should sit down to design a new system intact when all previous knowledge and design could be built on. Naturally, nobody did this in any literal sense before, but it appears important that the principle should be receiving emphasis. Thus from a 'scissors and paste' approach design one could proceed to a far higher level of design automation, or rather the speed and effectiveness of design automation. In other words, through technical advances the company is potentially able to circumvent the 'design-intensive' nature of their work in the past which had been an inevitable consequence of tightly constrained small-batch work. The extent to which they will capitalise on past designs through automation of the process is of course open to question. It does give them the option of perhaps splitting design engineering further by dividing it into elite groups working at a higher level than before and others who are merely pasting together past knowledge on something resembling a more routine footing. This change may underlie their attempt to bring the categories of designer and draughter closer together.

The engineering manager considered that organisation was far more important than mere technology in this process. Efficiency through the interplay of design automation and the 'captured' information of past designs could only be a reality if the organisation was right. Technology would achieve little unless it played a recognised and organised role in a company strategy. This picture should not be exaggerated. Even for the flat-plate aerials (which are the most extreme example of the trend) it is still considered desirable to have a drawing since if the design exists merely in the form of 'meaningless' numbers there will be greater possibilities of error, particularly when it comes to altering or repeating the design.

Also we were not able to assess the effect of new departures in product technology which, we were told, had, or were likely to have, far-reaching effects. This refers not to the advance in existing methods which could simplify work (ASICs for example) but complete reorientations such as the use of lasers in targeting. Whereas traditionally a radio (that is, sound) wave was sent to bounce off a target to get its position (on the same principle as used by bats), now the target is illuminated by light-waves.

To add to what was said earlier concerning captured information and its organisation, this is critically important in an organisation for which the vital 'new business' is, as often as not, the modification and repair of old business. The company cannot afford (M4 did for a time with new products) to survive on luck. Even if a design is not to be repeated, the information concerning it should be precise enough for repairs and redesign to be carried out twenty or even thirty years later.

Where drawings are useful in another aspect of the process is as the originals against which later repeats or actual items of equipment are automatically inspected. The inspection equipment used does not require an experienced quality control specialist. It consists of probes which, once their computer has digested the information from the original drawing, go out and do spot-checks and record errors.

As far as the drawing-office/design interface is concerned there has always been some confusion as to what is the division of labour between the two of these. With CAD, this confusion is concretised. In general, the principle followed at the company is that the designer decides the parameters, for example the face changer should be between the cells etc., and then the draughtsman goes and draws it. The designer therefore decides what is to be done and the draughtsman decides how it should be done. The onus of efficiency by design thus rests on the drawing office. There is currently an attempt to merge (or rather integrate) the two functions. In a sense they are divided as much by the traditional qualifications required for membership by the two groups as they are by technical/organisational logic. People in the drawing office will have ONC or HNC while designers will be graduates:

> At one time, the designer would go to the drawing office with the proverbial 'fag-packet sketch'. Now, the designer will receive questions and answers from the drawing-office. The designer's role becomes more theoretical. He will ask questions such as 'have you checked the minimum distance between signal parts?', and so on.

From the outside, it would appear that as it becomes more abstracted from the trial and error practices of traditional design the design engineer's job becomes increasingly susceptible to computerisation via expert systems or some kind of off-the-shelf stock of knowledge. The question then is whether the draughter has reoccupied some of this trial-and-error territory. Without having studied this as such, one is tempted to conclude that as the designer's work becomes more theoretical the draughter's becomes more routine since, as we have seen, there is an organisational preference for using stock solutions and capitalising on past design experience. What is interesting about the interface between the two groups is the way in which, depending on company priorities, both groups could move either towards or away from increased routinisation of their tasks.

In this case, the AUEW in the central machine-shop had successfully imposed a new technology ban for six months prior to our visit. No agreement was yet in sight. This did not mean that they would refuse to operate the eight flexible machine centres already *in situ*, but that they would observe a policy of 'this far and no further'. New equipment had arrived but was being left unused for the time being. For reasons that seemed difficult to ascertain, the senior managers interviewed did not seem overly concerned about the ban. Very likely the AUEW had reacted after the main objectives of management had already been realised. As the comparative restraint in CAD implementation (which had annoyed the MSF) suggests, there was a limit to the amount of automated manufacturing that was possible for the time being in terms of what products were required in the near future. As before we are reminded of the nature of the company as the large scale producer of small and very complex batches, something which led them to have a range of skills and equipment that would not always be in use.

The interface between the drawing office and the shop-floor is of interest. As we said earlier an unusual position has occurred whereby the MSF are demanding the extension of CAD which they see as enhancing their members' status and power, while management tell them they do not have enough business to justify such an extension, whilst on the other hand, the AUEW are attacking the MSF for participating in a shift–shift strategy to take skill from shop-floor to drawing office where the CNC tapes are prepared. The machine shop workers would prefer to programme the tapes themselves from drawings rather than have the tapes produced automatically via CAD/CAM links. In other words they see the management view of CAD/CAM as over-emphasising the CAD rather than the CAM. The computer-input potential on the shop-floor is scarcely utilised in many cases. In L3, a similar situation had occurred, only in that case it was a matter of CAD drawings being reprogrammed by CNC operators, in a process that amounted to duplication. Management there were trying to prevent the practice.

The role of management

Among the managers interviewed was a systems manager reporting only to the group general manager. The post has no subordinates and was created eighteen months ago to improve co-ordination and communication, particularly around, and in view of, the setting up of a general database for the whole company, including stores, accounting, and computer-aided engineering. There was a need for someone to take an overall view, without being hampered by vested interests, to bring all the various factions and departments together. The systems manager was doing this through an across-the board training initiative for supervisors and managers. The initiative was being run jointly with the training manager. CAE may be seen as the model and the core of the new system, since CAE involves this kind of shared database. In the case of CAE, the design engineer specifies a part or a design which is then recorded and called up whenever necessary by the CAD design draughtsman. An attempt is made to use each component for as many designs/uses possible, manipulating it in whatever way necessary. The specification links design with production, since it specifies which machine(s) is to produce the part concerned. Computerisation and CAE/CAD has led to the downfall of the drawing as a means of communication between design groups and production. What is now transmitted is a set of co-ordinates from one 'datum point', i.e. mathematically rather than visually. Although this leads to greater speed and accuracy overall, quite serious mistakes will obviously also be produced since it is more difficult to spot a mistake in a series of mathematical co-ordinates than on a drawing (which in the past the machine operator used to programme into the machine from the drawing that was passed to him).

All this has serious implications for the division of labour which have yet to be resolved, and which the systems manager is to some extent concerned with. If we take the design process as follows – the design engineer in the labs makes a specification and enters it into the computer where it is 'captured' for subsequent reuse whenever applicable by the design draughtsman on the CAD system (which is linked to and draws its information from the CAE computer) – it is clear that much of the content of practical work on the designs will centre on the modification of this captured information. Part X can be modified at the CAD end, having been called up from the database. It could also be modified at the CNC machine and, with the operator taking the basic co-ordinates and programming in the modification as he puts the tape into the machine for production. The managers interviewed considered that this was the arrangement that made most sense. The machine operator would make the necessary changes in the programme, run off one run of the part X and, if it were acceptable, would relay the updated version of the original programme back up to the drawing office, who in turn would enter the new variation in the database.

This conclusion should tie in neatly with the two managers' current concern that there was much downward communication and not enough upward communication in the company. They considered that a small batch skill-intensive company like theirs should not work in this centralised manner which is more fitting to a mass production company of similar size.

They are also aware that there is widespread resentment amongst shop-floor AEU members at the way the discretion has been removed from their level and moved up to the drawing office, with the encroachment of CAD into more and more of the product range. Even the job of inspecting the tapes for the CNC machines as they come from the drawing office has been taken over by a tape-inspecting machine. The drawing office has been successful in taking over the role of the white-collar production planning technicians who either did, or who supervised much of the NC machine programming in the past. Almost all of this group have been moved into the design labs where they provide technical back-up for the graduate design engineers. Encroachment by the new technology – CAD, and later its CAE link-up – has increased, having started on the most simple new products and parts first and then spread to the more complicated (this is the company approach with new technology). As the life-cycle of products is about five years, its introduction and its impact has been gradual, but now appears to have passed the critical points. Whereas it was once used for flat plates, it is now used for a wide variety of 3D products. The drawing office tapes are also fed into circuit simulation systems for checking and also go to automatic testing equipment (ATE). All this has meant the elimination of any major role for technicians operating between design and production (although the company makes sure that those operating the ATE etc. understand the product – some diagnostic skill is still necessary). This leaves the drawing office and the NC/CNC machine shops directly facing each other with a 'zero-sum' equation of skill between them – that is to say that every advance in skill, control or function by either means a corresponding decrease as far as the other is concerned. At present the sum is very much inclined in favour of the drawing office, while the machine-shop workers grow restive and are convinced that the company will soon have no use for them and has no respect or outlet for their skills. The managers are concerned about this, not least because they realise now, as we said before, that they will need trouble-shooting skills on the shop-floor to ensure efficiency, that there has been perhaps too much centralisation, and that they foresee an important role for modifications at the machine end (as was the practice in most German firms – see Sorge *et al.* 1983).

There is concern that the machine/programming skills may no longer exist on the shop-floor, since only the older workers have much experience of the type of progamming required, and their skills have decayed through lack of use. The manager sees the major barrier to change in this area as being the technology agreements they have signed with MSF on a national level, regarding the role of CAD draughtsmen in the design/production process. These agreements, which MSF members are profiting from, reinforce the draughtsmen's role, and the union is unwilling to change them and thereby concede skills and functions to the machine shop AEU members who would gain at their expense. Apart from anything else, the MSF might be worried that with their union now being more militant than their former partners in the AEU, that management might use any such revision of the agreements as the thin end of the wedge in squeezing the role of the CAD workers so that they became merely routine operators, losing all creative input in the process.

The managers were asked therefore what the management strategy was for work organisation around the design and production process (although much of it has become clear in this summary). The managers denied that 'management' had such a strategy, since such a word could not adequately describe the multitude of minor adjustments, negotiations and compromises involved. The systems manager in particular seemed very concerned with achieving compatibility between different groups both technically and 'politically'. Whilst a simple conception of the direction that technical change would move in could be a wheel around which all the different groups and stages of the process could be arranged in a way such that advances in one area 'knock on' to the others, in practice he said he found if difficult to balance the speed and nature of changes occurring simultaneously on opposite sides of the 'wheel'. Both managers always spoke of the CAD/CNC (or CAD/CAM) conflict as being a dispute between the MSF and AEU rather than between workers and management, although this might be seen as a convenient way of management to have the conflict for interest defined.

The hypothesis for what is actually going on could be this: the training manager is very keen on the survival of craft skills in the machine shop and elsewhere, and would thus like to decentralise in that direction. The systems manager wants consent for the changes being introduced and is thus concerned about the alienation of the craft workers. This makes both of them wish to emphasise the role to be played by craft workers in the 'modification at the machine rather than the CAD end' strategy. This policy does not mean that other groups within management do not prefer the CAD drawing office centralised approach to the process. What may of course disturb advocates of this strategy (and the training manager had previously said how impractical it would be to have the entire programming job taken out of the drawing office and placed back on the shop floor) is the extent to which it consolidates and extends the power of a politically militant union (whose enthusiasm for technological change in the design area becomes increasingly understandable). The draughtsmen are aware of this, hence their resistance to the delegation of any functions downwards, which could lay them open to further routinisation, losing out to the design engineers as well. More traditionally 'technocratic' management at Rangefinders no doubt believe the 'right first time' slogan of CAD/CAM, an idea which motivated the introduction of the technology at several other companies. The systems and training managers were openly sceptical about this. This further encourages them to decentralise the process. One possibility is that MSF draughtsmen might be squeezed into being a routine buffer between design engineers in the labs using CAE and a skilled trouble-shooting workforce on the shop-floor. MSF might conceivably prefer the elimination of the latter, with their members taking and keeping all trouble-shooting functions, and with any decentralisation of design work working in their favour; with increased discretion available to them, they would be able to use CAD's ability to remove the drudgery aspects of draughting as a means of improving their power, satisfaction and status.

In conclusion, we may observe how first, the complexity of modern technology defines simply-conceived automation and requires an (in the long run) increased reliance on skills at the lower end, compared to the experience of recent years. Second, the militancy of the drawing office/technicians' union, itself a result of the growth of white-collar militancy in the 1960s and 1970s, complicates the favoured management strategy (in Britain, though not in Europe) of concentrating skills and discretion above the shop-floor level. Third, in a large company there will be not one management strategy, but several, as change shakes up the *status quo* and potentially re-evaluates the position of different groups of managers, technicians, scientists and workers. In this kind of situation, traditional directive line-management ceases to effect strategy implementation or even formulation, which become the preserve of floating functional managers such as those interviewed. These managers, who operate without large departments under them (or without any departments at all), move about negotiating adjustments to existing arrangements in an attempt to minimise the traumas involved.

If this style of organisation replaces traditional line management, it would provide a buffer against a 'big-brother' approach to the centralised databases which are being set up as the complexity of product and process technology requires integration.

Concluding remarks

To conclude this review of the study findings, we may make the following points:

1 The drawing-office/machine-shop example demonstrates what is implicit in the rest of the above – that technology and skill requirements follow a pendulum course with each trend generating counter-trends which gather greater momentum and provoke contrary trends themselves. At each stage of the movement groups will exert their power to defend or advance their position in the process.
2 If management strategies tend to focus on key constraints on decision-making, shortages of 'off-the-shelf' graduates should provoke moves to reduce the requirement for graduate-intensive design and development.
3 The decay of skills within companies through lack of training, is as much a cause of the shortage as inadequate educational provision outside industry.
4 The flexible small-batch model (towards which many companies seemed to be moving nearer) both protects and erodes skills. While design intensity and the need for shop-floor intervention are emphasised, the complex task of repeating and modifying past designs leads in the end to greater automaticity and a concentration on 'captured' information.
5 Techniques such as CAD/CAM and CAE, whilst making flexible small-batch work as viable as was mass-production, may also make it resemble mass production in the long term.
6 This will be more likely in the large plants. In small firms flexibility and small batches meant greater skill-intensity and a less clearly defined division of labour.

In the Anglo-German CNC study cited earlier (see Sorge *et al.* 1983), it was suggested that parts become more complex so that the product as a whole can be kept within manageable limits of complexity. This process appears to have reversed where microelectronics have been applied. Under the continual pressure to customise, component units have been simplified so that more functions can be confined in a smaller space. As the newly available space is 'filled', we find an increasingly elaborate system with increasingly simple parts. The kind of design expertise is of a different order than before, when the needs of each complex part could be dealt with by a subject specialist and relatively easily linked together. Complex wholes require a different approach, with an emphasis on generalist and integrating capabilities, 'hybrid skills', and an awareness of the system as a product not as a scientific creation. According to Fores and Rey (1979), the conceptions of science and technology in English suffer from the lack of a concept analogous to the German 'Technik'. The latter concept emphasises and dignifies the creations of applied rather than pure science. Many of the problems we have recounted above could, at least in part, be accounted for by an over-clinical conception of the manufacturing process and an over-hierarchical and specialised conception of skill.

Regarding the model as set out in Figure 3.1 (see p. 29) we conclude that phases A, B and C are reflected in our findings and in the perceptions of those we interviewed. Phase D regarding 'hybrid skills' and an increased diversity of training is still in embryo in most companies although there are clear indications of the trend. Phase E indicating increased length and cost of training is as yet even less evident, although phenomena such as the 'elder statesman' training manager point to imminent upgrading for the training function, although still unsatisfactory increases in most training budgets. The last phase, phase F, involving decreased recruitment and employment, is obviously occurring in manufacturing although for reasons other than those we suggested it seems. Very probably the polarisation of the workforce between an elite and a larger periphery is a more profound economic and cultural phenomenon which will inform recruitment and training strategies rather than vice versa.

Lessons to be learnt?

The case studies outlined over the preceding pages suggest a number of overall similarities, which in turn have implications for management policy.

Use a centralised training strategy

As we indicated at the outset, all three cases are taken from very large companies. In each case, a vigorous training strategy was pursued from a department at group or divisional level rather than at company level. The training managers involved were thus able to operate with

Table 9.1 Summary of group case studies

	Aviation Ltd	*Alarms Ltd*	*Rangefinders Ltd*
Products	Aircraft systems	Alarms monitoring	Defence systems
Size of workforce	10,000	3,000	7,000
Product strategy	Vertical integration of systems design	Upgrading of product systems	Greater systems complexity
Skills requirements	Hybrid IT skills	Broader IT understanding	Broader up-to-date IT design skills
Skills problems	Graduate turnover and obsolescence of craft skills	General obsolescence	Graduate wastage
Training strategy	Training-up craft apprentices to graduate level	Basic modular training	High-level updating

proportionately greater resources than would be available at individual-site level, but without the bureaucratic 'distance' that might impede effective implementation from corporate level. In two out of the three, Training no longer reported to Personnel, while in the third, Training had come to dominate within the wider Personnel function.

Develop the leadership role of the training manager

In all three cases, a politically skilled manager with considerable commitment to the expansion of training activities had taken charge of a previously dormant training function. Each one had increased the resources available and attempted to raise the profile of training as a key strategic function. All had reacted sharply against the traditional Anglo-Saxon view of training as a 'soft' area, a welfare-related luxury to be cut back whenever any saving was required. Each training manager worked to maximise the cross-boundary influence of training, while exercising diplomacy to allay the fears of entrenched line management.

The centre-stage role of training was partly due to corporate perceptions of the need for cultural change, as well as a crisis over skills. (Both these factors derive their urgency from rapid technological advances in product technology.) The adaptation required cut across many areas of vested interest at all levels, hence the need for training managers who were able to 'out play the system', as well as possessing a long-term strategic HRM perspective.

'Target' resources as a response to anti-training attitudes

The strategies developed in each case were broad in scope, but formulated as a direct response to the challenges seen to be facing the company in question. Specifically targeted initiatives were built upon detailed analyses of the available manpower inside and outside of the organisation, and the mismatch between skills supply and the requirements of long-term company strategies regarding product and process technology. In none of the cases did a large increase in training resources proceed on the basis of training being an organisation panacea. Instead, all three training managers (in different ways) deliberately understated the extent of the resources expended in training, a consequence of having to fight for and justify each component of their budget in the face of parsimony on the part of top management and line management 'clients'.

The training managers actively worked to increase the impact of training through improved diagnosis and targeting, while minimising the resources perceived to be involved. Regarding the specific content of training initiatives, the following comparisons can be drawn:

Graduates

In two cases (Aviation Ltd and Rangefinders Ltd), the crucial skill-shortfall was seen to be at the 'top', with the lack of appropriately skilled graduates. In both cases, increasing the number of graduate recruits was not seen as a viable response. This view was linked to perceptions that the problem of graduate supply arose from higher graduate turnover (in Aviation Ltd) and high levels of 'stagnation' among the graduates who remained with the company (Rangefinders Ltd). In the first, the problem was being tackled through devoting resources to 'training-up' non-graduates (who were more likely to stay with the company) and then sponsoring them through university. In the second, an ambitious programme of graduate retraining with on-site university input was being carried through.

Technicians

Both Aviation Ltd and Rangefinders Ltd looked to the re-emergence of technician-level skills as an area of strategic attention. The maintenance of the apprenticeship schemes at both companies, and the emphasis on providing basic skills to craft trainees who would then 'convert' to technical apprenticeships reflected this view. Apart from this role, traditional craft training appeared to have a limited significance, reflecting the removal of discretion away from the shop-floor towards the drawing office as a result of the introduction of CAD/CAM.

Craft workers

A comparable decline in the role of craft skills was not evident in the case of Alarms Ltd. There, it was the extent to which these had fallen behind technical advances in product design that was seen to be the problem. The modular open-learning packages disseminated among the workforce were aimed at countering this disparity at a speed determined by the individual. The scheme was grounded in a gradual approach, with the emphasis on building-up commitment, of which little had previously existed in the industrial relations climate brought about by the insecurity associated with the skill disparity itself.

Management

Two types of management training appeared to be required. On the one hand, in Alarms Ltd, the 'anxious obsolescence' of a technically uninformed management was seen to contribute significantly to employee distrust as well as to management's inability to carry out its function effectively. Technical training for management (using the same packages as the shop-floor) thus had a significant cultural dimension. Rangefinders Ltd, on the other hand, was providing 'culturally oriented' management training in the strict sense. Supervisors and managers needed to be trained in how to manage a more skilled workforce in an organisation characterised by more flexible communication and less clear lines of control and demarcation. Management training also played a significant role at Aviation Ltd, although our interest is drawn more directly to the ways in which the Training Department attempted to educate other sections of management to accept their collective need for a longer-term HRM strategy for training cutting across departmental boundaries.

In all three companies, therefore, training managers were working to convince other managers of the need to develop skills of the groups of people already employed, rather than relying, as they had in the past, on the assumed existence of a limitless supply of already qualified recruits. Thus, an important lesson is 'development', as opposed to recruitment, where skills shortages are acute.

Additional points

Fear of change in the workplace, we would conclude, is the natural result of having vested interests in a particular *status quo*, whether regarding product technology, work organisation or even company culture. However adaptable someone who has spent years or even decades might be, the chances are that someone abler or merely younger and less steeped in the existing system may be in a position to profit more from the change. In the short term, it will be in the individual's interest to obstruct such change on these grounds. In the long run, however, the collective or corporate interest cannot be so served.

Training may thus be provided as a kind of 'organisational lubricant'. It may enable an easier adjustment of power bases so that, even if they are in many cases reduced, the long-term losses for those concerned will be cut. It also provides a means of more systematic selection for those who may hold greater bargaining power in the changed *status quo*. Training may thus be used to give such a shifting of the skill/power structure in a company the legitimacy of an egalitarian approach.

As a final note on the seriousness of the problem, we may recapitulate with the following observations: personnel and training strategies which had been predicated on the need for minimum skill-levels, produce generations of employees and managers unable to grasp the principles behind the processes with which they are dealing. When advances in technology lead to the integration of part of the whole of such processes, as has increasingly been the case, they are 'at a loss', and company performance suffers, along with individual or group security of employment. People who have 'graduated' through the impoverished educational spiral described above, cannot comprehend the complexity of the interfaces between the vast number of control mechanisms involved in the newer products. This fact leads to 'damage-limitation' strategies and 'piecemeal' technological solutions which have little long-term value. At the same time, the transmission of signals to and from remote end-terminals presents a problem for fault-diagnosis. Technical faults are not often such that they have their origin in places discrete from where they manifest themselves. In order to ascertain the origin of such problems, far higher levels of diagnostic skills are required than have become the norm in British industry. In one training manager's phrase, 'people who have so long been treated as idiots become idiots'.

Attempts to remedy this situation – particularly against the backdrop of the 'information society cult' we might argue – have tended in the view of Sorge *et al.* (1983), to concentrate on the assembly and communication of pure facts. This approach takes little account of the fact that many of those for whom such information is intended have no adequate 'frame of reference' in which to place these accumulated data. This may be even more true of some groups within management than it is for maintenance and other shop-floor groups. The opportunities for strategic choice thus seem to be particularly wide in the training area. The firms were sufficiently similar for one to have supposed that they might have arrived at similar training philosophies, although practices might have varied from one to the other.

Therefore, our conclusion might be to emphasise the options open to firms, rather than putting out pat solutions, least of all wholesale copying from either other companies or countries. The main recommendation is that training be given higher priority than before, and the organisational clout and resources to match, although this is contingent on the political skill of the training manager involved within the enterprise.

Organisational change and training strategy might then be seen as closely related, in the same way as industrial relations and training are integrated in the currently fashionable philosophy of human resources management.

We now turn, in the final chapter to some broader conclusions regarding the problem of training in British manufacturing, as compared with its German counterpart.

Chapter 10

Conclusions: the problem of training in British manufacturing*

The overall problem of skills shortage in the British economy may be seen as comprising two major aspects, both of which are detrimental to long-term competitiveness:

1 first, there is the large percentage of the labour force (49 per cent) who have no qualifications at all (Rose 1990b).
2 second, there are the skills 'bottle-necks' at the higher ends of both the academic and vocational training ladders (DTI 1985).

These two problem areas, to some extent complementary, would be serious enough in a stable context of technical development. However, the advent and diffusion (from the early 1980s onwards) of cross-sectoral technologies (microelectronics, biotechnology, optomatronics) have introduced a third problem area:

3 The phenomenon of 'technology shift'. If the core discipline of a product is gradually encroached upon and ultimately displaced by other disciplines, this exacerbates the effects of pre-existing skills shortages. Organisations in this context suffer loss of competitiveness through the lack of adaptability of their workforce whose skills have been insufficiently broad to cope with transition. Thus a short-term 'making do' approach to skills does in some sectors lead to a vicious circle of increasing shortages and longer development lead-times (Campbell and Warner 1989a).

TRAINING AND THE 'EXCLUDED MIDDLE'

In this chapter, it is argued that training policies and practices have suffered through insufficient attention being paid to the links between

*Part of this chapter is an amended *version* of a chapter published previously in J. Stevens and R. MacKay (eds) 'Training and Competitiveness' NEDO/Kogan Page, 1991. We gratefully acknowledge permission to reproduce it here.

these three problem areas. The 'fragmented' approach to training that has resulted has meant even less emphasis than before on intermediate craft and technical skills based on certificated vocational training. Rather than seeing the problems at each end of the skills market as symptomatic of too weak an intermediate skill base, decision-makers have tended to conclude the opposite.

They do not perceive the problem to be in the intermediate area at all, but that the focus of training strategy should be on the one hand the relieving of graduate skill shortages, and on the other the achievement of minimum standards of employability from what in Britain may now be termed the 'skills underclass'.

This tendency towards polarisation in the skills market is, it may be argued, a major source of inflexibility in British manufacturing. In so far as this is the case, it is ironic that it is legitimised as exactly the reverse – via the notion of 'core and periphery' labour markets. These are widely regarded as inherently flexible, although such a view effectively ignores problem area 3 above. If product disciplines merge and encroach on each other, shifting the firm away from its traditional base, the firm can only maintain control (of the courses of value-added) through a broad skill base which in turn facilitates a strategy of vertical integration of new product disciplines. In this connection, it is significant to note that both a propensity towards vertical integration and a broad skill base have traditionally characterised German manufacturing (Lawrence 1980, Campbell *et al.* 1989b). In Britain, by contrast, a simplistic version of the core and periphery approach to the labour market has increasingly been seen as an ideal to pursue, with the danger of producing an inflexible version of the so-called 'flexible firm'.

MANAGING 'EFFICIENT BOUNDARIES'

This is linked with a form of the 'markets and hierarchies' approach (Williamson 1975) under which managements avoid 'reinventing the wheel' by specialising in what they regard as their key source of competitive advantage, and put the rest out to contract. The markets and hierarchies approach suggests that the analysis of transaction costs should be applied as a means of determining where the efficient boundaries of the firm should lie. It requires situating the boundary of the firm at a particular point on the value-added chain (Reve 1990). The problem is not with the overall approach as such, but in its misapplication. The point is that value-added chains are not one-dimensional; each link of the chain serves as the focus of a variety of inputs, notably from different skill disciplines. A second complication is that the flow of work through the value chain is not unidirectional. As technology becomes more complex, design and redesign activities may occur on the shop-floor, and their results be communicated back to the design stage. It is paradoxical that, just as technologies such as CAD are being applied so as to effectively insulate design from production, so technological developments in products are, if anything, tending to blur the boundary between the two activities.

The argument is thus not that the 'efficient boundary' approach to the firm is inherently flawed, but rather that the decision-making related to it is likely to become more complex than advocates of the 'core skills, core processes' approach argue. This means that for the firm to function effectively it needs to take a broader view of what 'efficiency' in skill terms means.

If boundaries are in danger of being blurred, if the technical disciplines involved in the products and processes are altering in their relative importance to each other, then strong arguments exist for a more flexible base of skill at each stage of the value-chain, and for more overlap between different links in that chain. This does not rule out sub-contracting, which is likely to become increasingly prevalent. It does, however, mean that wholesale sub-contracting is not likely to be effective. A firm needs to maintain a broader activity-base and a broader skill-base than its immediate efficiency needs would dictate. In contrast, an entirely efficiency-led approach ignores the extent to which the bases of relative competitive advantage regularly alter, and how adaptability therefore becomes the central ingredient in long-term survival. The Anglo-Saxon approach to manufacturing has placed much emphasis on natural selection, but it has frequently ignored the extent to which competition on cost comes second to a general capacity for evolution as the basis for that selection.

In the preceding sections it has been argued that two popular notions, that of the core and periphery labour market and the specialised 'core skills and processes' firm, have, in their crudest form, legitimised severe structural weaknesses in the way skills are distributed amongst the labour force in Britain. In the following sections an attempt is made to demonstrate how these two (in our view) misperceptions of the skill context have affected both national policy and company strategies on skills.

SKILL FLOW NOT SKILL PROVISION

The skill shortgage problem is not a product of the 1980s but has been a recurrent problem throughout much of the post-war period. For much of the period, however, it was possible for the problem to be seen as a side-issue, since – and this is an important factor from an insular short-termist point of view – no 'proven link' existed between skill shortages and industrial performance (Meager 1986). Concern grew, however as international comparisons suggested a link between competitor countries' superior performance and their higher expenditure on training. The chorus of disquiet over British training reached a peak in the 1980s, notably with the publication of *A Challenge to Complacency* (Coopers and Lybrand 1985). It was also in the 1980s that training began to gain a higher political profile. However, whilst one may applaud the greater attention given to training by the governments of this period, it is argued that their policies have been characterised by some confusion over objectives. At a more fundamental level, there has been an over-literal espousal of the primary/secondary labour market theory, and the associated idea of the 'reserve army of labour'.

Thus, it was that some policies were aimed at facilitating the flow of labour, while others (in effect) worked to reduce it. The first tended to apply to low skills, the second to high skills. Regarding the first, the government, in the Wages Act of 1986, restricted the application of minimum wage standards set by Wages Councils of the Department of Employment in a number of low-paying sectors.

It was stated that the Act would improve job prospects for young people (Department of Employment 1986), although employers resisted the move, fearing that 'bad employers' would derive a competitive advantage (the minimum wage rates were after all designed as an absolute minimum, not a recommended level). Similarly, the social security reform of 1988 extended the qualification period for unemployment benefit from six weeks to six months if a job was left voluntarily. It had also earlier extended the qualification period for statutory redundancy payment from six months' to two years' service. All these measures appeared directed at establishing a 'secondary labour market' of low-skilled and low-paid workers, often in part-time work or in small firms in the service sector and in sub-contracting (Campbell 1988). It should be noted that the equating of small firms and secondary labour markets fails to hold true in the key high technology areas, where firms are required to pay high wages simply in order to attract and retain staff who are in short supply. Their low overheads may however mean that in areas such as software small firms have an advantage over large firms, and may 'poach' staff from the latter.

THE EMPHASIS ON THE EXTERNAL MARKET

The deregulation measures mentioned above, while having implications for the widening gap between the low-paid and skilled workers, between men and women and between small and large firms, had little relevance for the problems facing industry in terms of shortages of skilled labour. Much the same may be said of the trade union legislation.

The decline in trade union power which was associated with government legislation (both as cause and effect) did facilitate restructuring within firms and sectors, with the breaking down of demarcation barriers. However, as has often been commented, British industry as a whole remained too dependent on redundancies as a means of increasing productivity. A tailing-off of such increases therefore becomes inevitable, to the extent that these reductions in manpower were not sufficiently accompanied by capital investment. Perhaps to an even greater extent than with training, Britain's poor investment record has been camouflaged by the very short-termism and insularity which was its cause. Investment levels are assessed by comparison with the previous year, not by comparison with foreign competitors on a percentage of turnover basis. Trade union legislation thus encouraged an 'external' labour market approach to industrial change, and was not accompanied by sufficient emphasis on 'internal' mechanisms, apart from a (relatively slow) shift towards flexible working. In the same period the Redundancy Payments Act of 1965, originally designed to facilitate minor intersectoral adjustments, became perhaps the most important strategic instrument in the labour market. The flow of skills and labour away from unproductive plants was assured. What was not assured was the existence of new manufacturing capacity to absorb the labour concerned, and a new skill base to exploit this new capacity.

TEMPORARY WORK AND THE EXTERNAL MARKET

The pitfalls of too heavy a reliance on the external market are effectively illustrated by reference to the employment agency sector. In a study of fifteen employment agencies in 1987 (Campbell and Currie 1988), some concern was found to exist amongst technical staff agencies regarding their labour market role.

The agencies concerned were involved in moving craft and technical workers from one sector or region to the other on a short-term or temporary basis. To a greater extent than their clients, the agencies were aware of the real situation regarding skill supply. One spoke at length of the failings of British vocational training: 'We keep moving the same people around the country year after year, and they're getting older. No one is being trained. It's now almost impossible to find an ONC or HNC with two years' experience.' Another saw the agency role itself as ambiguous:

> The trouble is we're adding to the problem by giving clients an excuse not to train. They think they'll always be able to get people from us, whereas we know for a fact how few trained people are out there. We're now trying to encourage clients to train, because if they don't, we won't have any people left on our books in ten years' time.

Here one may perceive the irony of an approach excessively geared to external market measures. The external market agent par excellence, the employment agency, in the business of providing short-term solutions, finds that it is cushioning its clients from the real state of the labour market, and, in its own market interest, lobbies clients to take a longer-term view on skill provision.

THE 'PRISONER'S DILEMMA' OF TRAINING

The absolute decline experienced in many areas of British manufacturing raises the question as to whether market liberalisation alone can solve the structural problems of the economy, and of what mechanisms may be needed to complement the working of the external market.

This is not to say that the external market should in some way be restrained. Rather, the 'internal' market mechanisms (for example training with the firm) should be more than simply encouraged if the external market is not ultimately to fail. Here we may note Senker's (1988) observation that the market mechanism and employers' decisions are not one and the same thing. The decision may of course affect some aspect of the market's operation, but not its underlying principles. More importantly, if these principles are insufficiently understood, then the cumulative misreading of the market prevents that market from delivering what is desired, even though it continues to operate. In other words, if all employers expect to receive all their skills direct from the external market, then the freest market mechanism in the world will not deliver them.

It may of course be argued that 'the market' will correct itself – employers will realise that they have to train after all (and in this respect skill shortages do have some salutary effect). However, this self-rectification of the market is flawed by a version of the 'Prisoner's Dilemma'. If all, or even a large proportion of, employers conclude that training is in their interest, then there is no problem. If, however, only a few do, these few will become fewer as trained staff are poached away by non-training firms. Perhaps the most fundamental flaw in the 'self-correcting market' approach is that it ignores the parallel operation of supply and demand mechanisms regarding the price of skills. As the supply relative to demand of particular skills decreases, so the pay that may be expected by the holder of those skills increases. The extreme form of this trend emerged during the 1980s in the case of software engineers, of whom the elite 'the Computing Maradonas' carried what amounted to transfer fees.

If and when output expands during the 1990s, this may become true of quite 'mundane' skills, particularly where specific experience is required – for example HNC level in mechanical engineering combined with experience of the electronics industry, and vice versa.

What happens then, as supply relative to demand decreases, and relative wages increase, is that, rather than receiving an inducement to train, firms find it easier not to train at all. It might be expected that higher pay would encourage employers to train their own staff rather than recruit them in at exorbitant salaries. However the reverse may just as easily occur. Faced with the decision whether to invest in training people who may then leave for higher pay, a firm may decide that there is better value for money and less risk to be had by putting all the money into financing pay levels to attract staff from competitors. One company encountered in our research on skills pursued a policy of '10 per cent above the going rate and no training'. Low training investment thus begets even lower training investment. Shortages may encourage some firms to invest more in training, particularly if the skills that result are firm-specific and not certificated. However, just as many firms may see training as high-risk on account of poaching. They decide, rationally from the point of view of short-term individual interest, that the investment calculation favours the money being spent on higher wages to attract people in, rather than training to raise the skills of those already inside.

A LONGER-TERM APPROACH FOR MANAGEMENT

These tendencies have exacerbated what would in any case have been a severe shortfall of graduates. They have also increased graduate dissatisfaction (graduates having been given work that did not meet their expectations) and inflated graduate turnover in companies where the shortage of graduate recruits had already become a major problem.

In one of our cases, the training manager was able to demonstrate to senior management that the high level of turnover (60 per cent within the first year of service) had inflated the costs of graduate recruitment and induction to the extent that they were no longer justified by the small numbers who actually stayed. Whereas sponsorship of existing employees on courses up to graduate level had previously been dismissed as too expensive, this could now be shown to be cheaper in the long term on the basis that existing employees, being locally based and more entrenched in the firm, were far more likely to stay in the long term. It is interesting to note that although figures could be produced to demonstrate that sponsored employee graduates were indeed less likely to leave, the short-term orientation of senior accountants meant that they persisted in regarding recruitment as cheaper than training and sponsorship.

A large amount of 'political' lobbying was necessary for the change in strategy to take place. The training strategy set in motion by the training manager was a relatively ambitious one. The idea was to take shop-floor workers who had been de-skilled through automation and train them up to graduate and even postgraduate level. The elite of each group was selected for training and sponsorship (ONC, HNC, as well as degrees) to fill out the skill gaps in the next level up. Thus over a period of years, the organisation would have evolved a reverse 'cascade' of skill provision increasingly independent of recruitment and protected against the movements in the labour market owing to the already existing loyalty of employees to the firm and its locality.

Although senior management had come to accept the training/sponsorship argument against relying on recruitment, they were not prepared to meet the investment costs of the programme the training manager had outlined. Although more expensive than recruitnment in the long term, the training programme would require more money 'up front' rather than spread across successive financial years as recruitment costs had been (hence the accountants' view that recruitment was 'cheaper'). However, this did not seem to prevent the training programme from going ahead more or less as planned. The reason for this was that the training budget was more flexible than might have appeared.

The training manager explained that in the culture of this company (which was not seen as markedly different from other large British companies), the fact that money was seen as a cost rather than an investment, and therefore needed to be seen to be kept low, did not mean that the money was not being spent. Initiation into management was seen to involve the recognition of two areas of ambiguity. First, budgets could be manipulated since they were merely a representation of reality rather than the reality itself. Second, rebukes from senior managers were not to be taken literally, but were often a matter of ritual and appearances.

TWO LEVELS OF PARTICIPATION

As Senker (1988) has argued, the choice is between:

1 encouraging employers to train for their own needs, in which case the emphasis must be on retention of staff and a less dynamic market for skill, or
2 encouraging employers to draw skills from the market, in which firms need to be encouraged to train beyond their own needs to supply skills to the marketplace.

Participation is central to both approaches, although a different type of participation would apply to each case. In the case of 1, the strategy would need to be grounded on participation within the enterprise with systematic consultation on day-to-day issues. In the case of 2, a system of training levels would be required, and the model for achieving this would be an improved version of the tripartite 'corporatist' training boards.

Thus, there is a choice between firm-based training and sector-based training. The option, in turn, entails a choice between enterprise-level participation and sector-based mechanisms such as compulsory levies to assist in making the operation of internal and external labour markets a mutually reinforcing one, rather than the vicious circle that has existed in the past. A clear policy is essential if the middle level of skill is not to be further eroded, leading to further exacerbation of the under- and over-supply problems existing at the high and low skill levels.

COMPARISONS WITH THE FEDERAL REPUBLIC OF GERMANY

Detailed comparisons between the British cases from this study and those of the parallel German study have been made elsewhere (Campbell *et al.* 1989a, 1989b). To develop the argument outlined there, we will summarise the comparison around two themes only: (1) recruitment and training and (2) product strategy. In doing so we will emphasise the relationship between the two. In particular, we will emphasise the possible dangers of focusing on one 'core' part of the product and ignoring the extent to which technological advances may shift the 'core' or key area of value-added, from one part of the product to another, to conclude the book.

Main findings 1: recruitment and training

The overall findings may be divided into two types for the purposes of this review. On the one hand, there are findings related to the types of training and recruitment carried out by the companies in the two samples. On the other there are the more fundamental differences in product and manufacturing strategy which provide the context for the personnel practices involved. It is argued that differences in the second area may have a greater long-term significance than those in the former category where, albeit slowly, some convergence was beginning to appear (that is, if one allows for the most significant difference of all, the lack of a coherent national system of vocational training in Britain).

Regarding training practices, there was some confirmation of the Anglo-German differences already familiar from the work of Prais and Wagner (1983) and Sorge and Warner (1986) amongst others, although a confirmation at least partly mitigated by a degree of convergence in some key areas of difference. British companies, like German companies, had been developing closer links with public education, there had been a degree of internal 'training up', although not on a scale comparable for the mid-career technical training of craft workers in Germany, and there had also been an expansion of training (technical and attitudinal) for employees in supervisory positions.

These tentative signs of convergence in individual companies should not mask the most significant difference – that is, the reliance in Britain on a voluntarist or individualist system of vocational training as opposed to the more collectivist dual system of the Germans (see Keep 1989a). Thus, any convergence from the British side tended to be the result of individual initiatives in particular firms which, although impressive in themselves, did not appear to present an adequate solution to problems at sectoral or national level.

If the main differences remain, albeit in a partially mitigated form, there was a clear convergence in terms of the problems perceived by respondents in the two countries. In both the UK and West Germany, skill shortages in terms of electronics hardware and software graduates (shortages regarding both quality and quantity) were seen as the most pressing human resources problem in a majority of the firms, and in large firms in particular. There was little to distinguish between the individual responses of companies in the two countries to this problem; in both there were attempts at conversion training for graduates from other disciplines, and in both there were a variety of methods (more selective recruitment, some extra training) applied to reduce the problems caused by hardware and software specialist recruits not being acquainted with the needs and disciplines of the industrial sector in which they were now employed.

Main findings 2: strategic and structural differences

A survey by Northcott (1986) had found that the gap between the rates of diffusion of microelectronics product applications in British electrical/electronics engineering sectors was wider than that occurring between comparable companies in West Germany. German mechanical engineering firms were seemingly more able to incorporate technical shifts in the product than their British managerial counterparts.

This finding was echoed in our own study described in earlier chapters. In our sample, all the firms covered had, by definition, incorporated microelectronics into at least some of their products. It was found, however, that the British mechanical engineering firms which had incorporated microelectronics were more dependent on outside suppliers for microelectronics-related development and production, than were their German counterparts who had for the most part successfully vertically integrated regarding microelectronics. This is described in more detail below.

Cases from the two countries were classified according to whether microelectronics affected a primary or secondary product function, i.e. whether or not information processing was what the product was primarily used for. Where information processing was a primary function (as in the electronics industry generally) such companies were placed in category A. Where information processing was a secondary function (as in mechanical engineering or aeronautics), such companies were placed in category B. A and B type companies were then compared to see how far they had vertically integrated microelectronics-related development and production.

This aspect of the findings was important in that the greater the vertical integration, the greater the demand for microelectronics-related skills. Conversely, the fewer the skills available, the less a strategy of vertical integration would be practicable.

It was to be expected that in both countries, A type companies would have vertically integrated microelectronics to a greater extent than B type companies, since in the latter cases information processing (and therefore microelectronics) was not the main focus of the products. What was less expected was the relatively clear-cut pattern of national difference that emerged from the sample. A large majority of the German companies appeared to have vertically integrated microelectronics with many companies in both A and B categories being involved in the development and assembly of microelectronics sections of the product down to component level. Although A companies in the German sample went further than B companies on the whole, the separation between the two was not as wide as had been expected.

In the British sample, by contrast, not only was there less vertical integration overall, but also a much more marked polarisation between A and B companies. Almost all the non-information processing companies were wholly dependent on outside suppliers for development and assembly of microelectronics-related sections of the product, whilst the larger A type companies were on the whole attempting to move away from direct involvement in component-level development and assembly towards a more systems related approach to the product, where value-added lay more in the arrangement and linking of modules rather than in the design and complexity of modules themselves. These findings regarding manufacturing strategy are presented in detail in Campbell *et al.* (1989a).

Implications of product strategy findings: a benign view

On the surface, the findings for the British sample in terms of product and manufacturing strategy give no cause for alarm. They could, on the contrary, be taken as suggesting a plausible trajectory of successful industrial adjustment. According to this view, expertise associated with microelectronics hardware and software would be concentrated in larger electronics-based firms (who in turn could draw on software houses and consultancies). As both hardware components, sub-assemblies and software became increasingly standardised into off-the-shelf packages, and application-specific chips removed the need for more sophisticated hardware design (and there is evidence to support this part of the logic), these firms would sharpen their make/buy decision-making. They would pull out of direct intervention below module level, and concentrate on what they did best, and therefore avoid reinventing the wheel' (an almost unanimously expressed fear in the British sample). The main thrust of training and recruitment policy would then be to ensure an adequate supply of cross-disciplinary systems engineers, capable of operating across the different modules of the electronic system and its links to the other sections of the product. This approach would form part of the trend towards multi-firm projects, with the specialist inputs of different firms co-ordinated by generalist project managers who had an awareness of the product as a whole.

We would thus find a complementary relationship between electronics/information processing and mechanical/non-information processing firms. The first type of firms would hold the design authority for the microelectronics-related sections of the products produced by non-electronics-based engineering firms, who would therefore have little need to develop their skill base in relation to microelectronics. In return, the non-information processing firm would keep the electronics/software supplier aware of the needs of the product so that the information processing aspects served its function (and customers) appropriately.

The plausibility of the above is such that the relative absence of evidence for it in the German sample comes as something of a surprise. One explanation that may be offered is that whereby the strengths of the German system in the post-war period – high status for production, emphasis on quality, delivery and detail, and therefore also on practical, intermediate skills (see Lawrence 1980) – may ultimately become its weaknesses. A stolid German thoroughness would, in an age of low-cost standardised hardware and software, become obsolete. Attention would be shifted towards the more 'noble' aspects of manufacturing (ironically the term was cited by German aerospace managers), namely the cerebral orchestration of the ready-made parts into systems. This shift would then justify the 'flight of skills from the factory floor' in British manufacturing (Sorge *et al.* 1983, Senker and Beesley 1986) after all. Although there are elements of truth in the scenarios outlined above, they mask serious problems in the British case, just as the 'post-industrial society' paradigm, very popular in the early 1980s and which suggested a shift towards information skills, helped to camouflage and justify what was in fact a simple case of British manufacturing failure (Hampden-Turner 1984).

Implications of product strategy findings: a less benign view

It was stated earlier that British and German firms suffer from similar skill shortages. The shortages operate in different industrial contexts, however. Although it would be an exaggeration to state that the shortages in the one were a function of sectoral weakness and in the other a function of strength, it was found that German mechanical engineering firms have microelectronics-related skill shortages because they have incorporated microelectronics-related development and assembly into their in-house activities, whereas large British mechanical engineering firms have shortages because they have not been able to do so and the shortage continues to prevent them from doing so. In one British case, for example, shortages of electronics hardware and software engineers occurred because the job assigned to them, that of quality assurance of work done by suppliers (who have full design authority over the electronics side of the product), is insufficiently interesting to retain people who are in demand elsewhere. Such findings are not necessarily an indication of the weakness of the firms themselves, but rather of the relative weakness of engineering firms as recruiters in the British economy. The high status of the financial and retailing sectors in the economy has meant that engineering firms have difficulty in attracting and retaining people qualified in disciplines which are also in demand in those sectors (as has been the case with electronics hardware and software over the last ten years).

The role of the competence shift

The main point we must underscore here is that the West German mechanical engineering firms appear to have coped better with the competence shift occurring in the sector. In doing this they have been hindered by skill shortages at graduate level, but have been able to fall back on the relative strength of their broadly trained (and retrained) intermediate skill-base. The previous 'benign' scenario suggested that mechanical firms could rely on electronics firms to provide those parts of the product which now involve microelectronics. In practice, not only is it difficult to fulfil this liaison effectively without having a critical mass of the relevant skill in-house (so that the output of the different disciplines can be tailored to the needs of the product as a whole), but also the disciplinary shift is not a one-off occurrence but a continuing trend. Thus, in one complex mechanically based product, the proportion of total value accounted for by microelectronics based controls had increased from 15 per cent to 25 per cent over five years, while in aircraft manufacture it was said that the value of electronic control systems would eventually surpass that of all other components within a few years. If and when this occurred, it was not certain that the airframe manufacturers would remain the prime contractors, but could instead become the sub-contractors of electronics/information processing companies.

Although this shift had been under way in the days of electro-mechanical controls, miniaturisation of circuitry had accelerated it considerably. With each stage of miniaturisation existing functions would be compressed into a smaller space within the product, and provided at a lower price. Customer pressure (which in most firms represented a more consistent influence on product development than before) would then demand more functions to be added in to 'fill out the space'. In some relatively simple products, a 'ceiling' is soon reached whereby no further additions of microelectronics-based functions are required. In more sophisticated products, however, this 'ceiling' continually recedes; more and more microelectronics-based functions are added in, their total value increasing even as their unit cost falls, as we have seen. In terms of the core competence changes outlined above, therefore, it can be seen that there is a significant Anglo-German difference in the way industrial change is perceived. The picture suggested by the British sample is that of the subordination of mechanical engineering firms to electronics firms, or, worse, a 'shake-out' of mechanical firms and a complementary growth (as yet not as significant as expected) in the electronics sector. From the German sample, the picture is that of industrial transition within the enterprise itself whereby new disciplines are incorporatred into the core competences of the firm, this shift being accomplished through retraining.

Design and production

Differences still appear even where British companies have managed to make this transition. Several British companies in the medium-sized bracket had achieved a dramatic shift in the weighting of disciplines from being wholly mechanical, to being between one-third and two-thirds electronics-based. Where this has occurred, however, it has frequently been accompanied by the intensification of another characteristic of British industry which is relatively under-represented in Germany, namely the separation of design from production. In one case, a mechanically based firm had built up an electronics design and development facility which had gradually become relatively independent, increasingly taking on its own contracts from outside, without any involvement from the overwhelmingly semi-skilled mechanical workforce. In other firms, design and production took place on wholly separate sites, with a minimisation of skill not only at the production end but also in the prototype building stage of design, so that 'our too many clever people down there' would not improve on the designs once they left the CAD drawing office (such improvements would not be recorded officially and would therefore interfere with inter-site communications).

Considerations such as these may underlie the relative lack of a skilled electronics workforce (or a hybrid mechanical/electronics workforce for that matter) in the British sample. The assumption was that technical change meant a shift from blue collar to white collar, from production to design, and from mechanical to electronic.

One large firm which had begun the competence shift at an earlier stage (most of its management were former mechanical engineers, while most of its engineers were in electronics hardware) was accelerating it through transferring apprenticeships (and individual apprentices) from mechanical craft apprenticeships to technical electronics apprenticeships. This reflected the established British tradition of 'early extraction of elites' (Sorge and Warner 1986), in effect an attempt to relieve skill shortages at higher levels by draining the lower levels of the more talented workers. This would set a vicious circle in motion whereby the pool of intermediate skills would be dried up, and no remaining craft workers would be of sufficient standard for retraining or cross-training at a later stage, thus increasing the already existing overemphasis on recruiting already qualified staff at technician or graduate level.

This observation ties in with a tendency in large British engineering firms, when implementing process innovations such as CAD/CAM, to see that the discretion threshold, the level in the skills hierarchy where active intervention in the process can occur, is kept as high as possible (Campbell *et al*, 1989b). Such interpretations of new technology by management reflect management's impatience with shop-floor demarcations, through which have been preserved the narrow conceptions of training and skill that have characterised British industry (Keep 1989b).

Managements in Britain (despite much rhetoric in recent years) have found it more convenient to undercut workforce skills, rather than develop them towards flexibility. This state of affairs contrasts with the position in West Germany where, at least in core industries, German management 'is primarily concerned to obtain functional flexibility and is not motivated by attempts to achieve a downward adjustment of terms of employment' (Lane 1988).

'Information society' and the pitfalls of elitism

Lack of awareness regarding the strategic importance of training has been endemic in Britain at senior management level, and has been the subject of much criticism, especially since the Coopers and Lybrand's report of 1985. Equally responsible, however, are the policy-makers closer to government who have, in the earlier part of the decade popularised the idea of de-industrialisation as if it were a form of progress (see Pollard 1982, Williams *et al.* 1989).

At the level of smaller companies, or operational levels of larger companies, there has been an increasing disillusion with what might be termed the 'information society' approach to technological change and skills (see Sorge *et al.* 1983). This view suggested a strategy of vertical dis-integration, with the core of the firm becoming the preserve of 'information workers'. In practice, managers in both the UK and West Germany spoke of their increasing disinclination to recruit graduates or technicians with purely software-based qualifications. These were seen as not only causing problems through upsetting pay differentials as a result of their market scarcity, but also being generally unaware of the nature of the product. However, moving away from pure software qualifications (to qualifications of mixed hardware and software, or software in relation to a particular industrial sector) appeared to be an easier option in Germany where electronics degrees were said to involve hard and software in most cases. This mix contrasted with Britain, where the prevailing enthusiasm for 'information' skills in the early 1980s had led to a disproportionate increase in the numbers of purely software engineers available (although, paradoxically there were still too few to meet the cross-sectoral demand).

In more complex products, the need for hybrid skills, and for a broader view of the product as a whole, had become critical. In many cases the boundaries between the contributions of different disciplines had blurred, so that the whole became considerably more complex than the sum of its parts. For at least one respondent in the British sample, the difficulty engineering firms had in coping with these developments had been heightened by the tradition of over-specialisation, not only at shop-floor level but also at the level of graduate designers. In this training manager's view there had been a reversal of the skill requirements trend of recent years; rather than there being a need for more highly qualified specialists, there was in fact a need for more broadly practical intermediately qualified staff ('such as a good ONC') who, in his view, would be better placed to grasp how systems were linked together.

With the drastic cut-back in apprenticeships, such people were often more difficult to recruit than graduates (although less difficult to retain).

The shortage of intermediate skills may be placed in the context of the dis-integration strategies referred to earlier. A number of companies seemed to approach this strategy as if it was synonymous with a university-trained specialist core and a periphery of sub-contractors and occasional agency staff. In fact, it could be argued, the strategy stood more chance of being successful if it was founded on a base of generalist skills at the craft/technical level. British companies in the larger category tended to hold an elitist view of skill requirements, assuming that graduate-level qualifications were needed for jobs which could, in fact have been carried out by technicians or even craft-level workers, given some extra training. There was indeed a need for more generalists at all levels of skill, not just the highest.

Markets and hierarchies

The strategies of the British companies as a whole appeared to follow a version of the 'markets and hierarchies' model (see Williamson 1975), perceiving the graduate-oriented design function as the core and other manufacturing activities as the periphery. This meant keeping control of that which was most difficult to acquire on the market, and buying the rest in. This view is linked to that whereby each firm has a 'core competence' which it should regard as its key source of value-added (Francis 1988), and that time spent co-ordinating the rest of the process would therefore be a waste of resources.

This view may be faulted on three counts:

First, it ignores the extent to which, as we have tried to demonstrate, core competences and disciplines are themselves shifted through the application of new technology in products.

Second, it focuses attention on a misleadingly elitist conception of human resources rather than on the product itself. What we have attempted to argue in this section is that for a more systematic control over product quality (both in design and manufacture), people with generalist skills and experience at the intermediate level of qualification (or trained up from an intermediate level) provide the best basic skill base, with which graduate specialists could interact more effectively than if left to themselves. Indeed, product quality considerations had encouraged one of the smaller high technology firms in the British sample to bring production back in-house, having previously sub-contracted all of it.

Third, the shortage of graduates has led to a 'sellers' market' in graduate recruitment, and correspondingly high rates of turnover. These have in turn increased graduate recruitment costs. Several companies in the sample were finding it better value to train people from the shop-floor level to technician level, and then to sponsor them through university. The graduates who were produced from this internally driven process were found to be more likely to stay with the employer. The viability of this strategy depends on talent being recognised and developed at lower levels within the firm. If a graduate-oriented core and periphery model is adopted, the possibility for this solution to high-level skill shortages is drastically reduced in the long term. The result, ironically enough, is that the graduate-oriented firm then succumbs to the first serious wave of graduate skill shortages, whereas it would not if it had an intermediate skill base available for training up or sponsorship.

Manufacturing management strategy and intermediate skills

To sum up the argument thus far:

1 British manufacturing strategies reflect a concept of dis-integration which overemphasises specialist skills and underemphasises the importance of the product as a whole.
2 They are reinforced in this model by a lack of the intermediate skills which could suggest a different route.
3 This reinforcement occurs not least as the result of a traditional fear of reliance on shop-floor skills (or even technician-level skills, where this can be avoided). It was not unusual, in the British sample, to be told that 'we're still more dependent on shop-floor skills than we would like'.
4 The lack of a generalist skill base inhibits rapid shifts of competence, when these are required. It has been the case with microelectronics where many companies are concerned, and is likely to be the case with the diffusion of optomatronics (the combination of technologies of light with mechanical and electronics technologies).

5 This position contrasts with the German case where there has been a continuing emphasis on taking control of new technological disciplines in-house, and ensuring that at least some of the detailed design and assembly work in new areas is carried out by the firm's own employees.
6 Such a tendency is not unconnected with the relative strength of the intermediate skill base in German firms, maintained through sector-wide collective agreements dealing with the extent, funding and quality of training and retraining activities.

Further national differences

German companies were less prone to functional differentiation in management – there was greater overlap between the marketing and technical functions and the training and development associated with them (although systematic liaison between engineering and marketing was seen to have increased markedly in a number of British firms). In British companies, career and organisational splits were more frequent and pervasive than in the German companies. Technical and project management career paths in engineering were divided at an early stage. In one British company this was seen as a means of 'keeping boffins out of harm's way'.

In Germany, even in large firms it was more common to find technicians with relatively modest qualifications working as hardware designers. This was rare in Britain, where graduates showed a marked tendency to want graduates to work under them.

Craft-level skills in electronics saw the greatest contrast. West Germany provided twice as many electronics craftsmen via examination as Britain did. British firms were more in favour of technician-level electronics apprenticeships. This may be connected to the observation of Prais and Wagner (1983) that German electronics apprenticeship examinations were more demanding. It is possible that British technical electronics apprenticeships were more comparable to the craft electronics apprenticeships in Germany. In several of the British firms, craft electronics work appeared to be confined to conceptually simple tasks. In Germany, on the other hand, half the sites investigated provided electrical and electronics worker apprenticeships, which provided the basis for work in different functions and possible further adult education for technician and engineering posts. Technical training off the job is important for supervisory promotion, unlike with the 'separate hierarchies' principle in Britain. Rather than maintaining production in its role of 'pariah', still frequently the case in Britain, the German system appears to link production with 'higher' functions.

The unofficial training budget

One unexpected national difference was the practice, found in more than one large British company, of building up an 'unofficial' training budget. This represented an extreme (but apparently necessary) corollary of the voluntarist system, whereby training managers would find ways of extracting money from client departments, for example by loaning apprentice labour to projects in return for a percentage of the project's success. This avoided lengthy rituals of financial justification with senior management. Such informality, whilst not officially sanctioned, was said none the less to be expected by senior management as part of the 'rules of the game'. It was not that training expenditure had to be kept low in practice, it merely had to be seen to be kept low. At least three large sites did spend between 2 and 3 per cent according to this system, while their official figures were under 1 per cent. The unofficial accumulation of training resources was facilitated by the market relation of training to other departments in the British companies, a system less common in Germany. One of the training managers concerned saw the system as necessary on account of senior management being out of touch with technical developments and not knowing what their training response should be. Whilst not admitting this to be the case, they would give those who were better informed an (unofficially) free hand.

The German sample presented few examples of such initiative, or of such informality. There were on the contrary some signs of a lack of flexibility in company training strategies for non-graduate employees. Although the vocational training system in Germany has the advantage of a clear progression between the distinct stages of craft, technical and supervisory levels, none of which can normally be omitted, this apparently, in some cases at least, left little scope for technology-related further training for craft workers if this was not linked to formal promotional training. Additionally, there was almost as little training for semi-skilled workers in German firms as there was in Britain.

We would still maintain that these weaknesses do not (as yet) significantly detract from the advantage which the intermediate skill base with its broad apprenticeship structure gives German firms of all sizes when coping with the demands of technological and competence change.

Conclusion

It is tempting to reiterate the relatively dismal record of British industry on training, and the extent to which Germany, among other European countries, has evolved a system better equipped to evolve new 'production and employment concepts' (Kern and Schumann 1987) in order to cope with rapid technological changes in products and processes, increased customer expectations and tighter competition on non-price areas such as quality and delivery. In this book, which echoes these criticisms, we have given some attention to convergences in British and German practices, while emphasising the considerable structural differences that have arisen as a result of different manufacturing (and skills) strategies. A number of examples were none the less found where results analogous to those achieved by the national vocational system in Germany had been achieved through the localised and sometimes covert strategies of individual managers in British companies.

There is little doubt as to the areas in which British skills provision strategies need to be improved. What is more problematic is the means by which this improvement should best be achieved. In some respects, British training is torn between two courses of action. On the one hand, taking 'success stories' as a model, the problems of training could be left to be resolved through individual initiatives, thus following the voluntarist tradition. This approach rests on the assumption that broad-based education and awareness at senior management level will percolate through into training strategies that will contrast with the neglect or narrow specialisation of the past. Given that it is the voluntarist tradition itself that lies behind the present skills crisis, this may prove too risky an assumption.

Bibliography

Albu, A. (1980) 'Changing Attitudes to British Engineering Education', in K. Pavitt, (ed.) *Technical Innovation and British Economic Performance*, London: Macmillan.

Armstrong, P. (1987) 'The Abandonment of Productive Intervention in Management Education Syllabi', Warwick Papers in Management, Warwick Business School.

Bailyn, L. (1987) 'Experiencing Technical Work: A Comparison of Male and Female Engineers', *Human Relations* 40 (5), 299–312.

Barnard, C.I. (1938) *The Functions of the Executive*, Cambridge, Mass.: Harvard University Press.

Bell, D.A. (1984) *Employment in the Age of Drastic Change: The Future with Robots*, Tunbridge Wells: Abacus Press.

Benton, L., Bailey, T., Noyelle, T. and Stanback, T. (1991) 'Employee Training and US Competitiveness'.

Beresford, P. (1986) 'Getting it right in the shires', *Sunday Times* (Supplement), 26 January p. 22.

Blackburn, P. and Sharpe, R. (1988) *Britain's Industrial Renaissance?* London: Comedia/Routledge.

Braverman, H. (1974) *Labor and Monopoly Capital*, New York: Monthly Review Press.

Buchanan, D. (1983) 'Technological Imperatives and Strategic Choice', in G. Winch, (ed.), *Information Technology in Manufacturing Processes*, London: Rossendale.

——(1986), 'Management Objectives in Technical Change', in D. Kings and H. Willmott (eds), *Managing the Labour Process*, London: Gower.

Campbell, A. (1985) 'New Technology and Management in the British Coal Industry', Unpublished Ph.D. thesis, Henley, The Management College, and Brunel University.

——(1988) 'The Case of Great Britain', in P. Auer and H. Fehr-Duda (eds), *Industrial Relations in Small and Medium Sized Enterprises*, Berlin: Instituut fur Angewadte Sozial und Wirtschaftwissenschaft (IAS).

——and Currie, W. (1988) 'The Nervous Industry: Ambiguities in the Labour Market Role of Private Employment Agencies', Paper presented at the

Annual Conference on the Organisation and Control of the Labour Process, Aston University, March.

——Sorge, A. and Warner, M. (1989a) *Microelectronics Product Applications in Great Britain and West Germany*, Aldershot: Gower.

——Currie, W. and Warner, M. (1989b) 'Innovation, Skills and Training: Microelectronics and Manpower in the United Kingdom and West Germany', in P. Hirst and J. Zeitlin (eds), *Reversing Industrial Decline?*, Oxford: Berg.

——Sorge, A. and Warner, M. (1990) 'Product Strategies and Human Resources: Anglo-German Differences', *Journal of General Management* 15 (3), 39–54.

——and Warner, M. (1986a) 'Innovations, Skills and Training: Microelectronics and Manpower in Britain and Germany', Paper presented at Conference on Industrial Structure and Industrial Policy, Birbeck College, London, 14–16 July.

——and Warner, M. (1986b), 'Production Innovation, Skill Needs and Manpower Training: A Study of Microelectronics Applications in Selected British Companies'. Proceedings of the Third International Conference on Human Factors in Manufacturing, Stratford-upon-Avon, 4–6 November.

——and Warner, M. (1987a) 'Microelectronics Technology, Industrial Change and Work-Organisation: Towards a Model of Skills-Training', Working Paper, Henley, The Management College.

——and Warner, M. (1987b) 'New Technology, Innovation and Training: An Empirical Study of Selected British Firms', *New Technology, Work and Employment* 2(2) 86–99.

——and Warner, M. (1987c) 'New Technology, Innovation and Training: A Survey of British Firms', *New Technology, Work and Employment* 1 (2), 86–99.

——and Warner, M. (1989a) 'Training Strategies and Microelectronics in the Engineering Industries of Britain and West Germany', Paper presented at the Training Agency Conference on International Comparisons of Vocational Education and Training for Intermediate Skills, Manchester, September.

——and Warner, M. (1989b) 'Organisation for New Forms of Manufacturing Operations', in R. Wild (ed.) *International Handbook of Production and Operations Management*, London: Cassell.

——and Warner, M. (1990) 'Managing Advanced Manufacturing Technology', in M. Warner, W. Wobbe and P. Brodner (eds), *New Technology and Manufacturing Management*, London: Wiley.

Child, J. (1972) 'Organisation Structure, Behaviour and Performance: The Role of Strategic Choice', *Sociology* 6 (1), 1–22.

——(1984) 'New Technology and Developments in Management Organization', *Omega: International Journal of Management Science* 112 (3), 213.

Clegg, S. (1990) *Modern Organizations*, London: Sage.

——and Dunkerley, D. (1980) *Organization, Class and Control*, London: Routledge & Kegan Paul.

Connor, H. and Pearson, R. (1986) *Information Technology Manpower into the 1990s*, Brighton: Institute of Manpower Studies.

Coopers & Lybrand (1985) *A Challenge to Complacency*, London: MSC.

Cross, M. (1985) *The Flexible Craftsman*, London: The Technical Change Centre.

Department of Employment (1986) *The Wages Act, 1986*, London: DE.

Department of Trade and Industry (DTI) (1984) *The Human Factor: The Supply Side* (First Report), (1985) *Changing Technology, Changing Skills* (Second Report), IT Skills and Shortages Committee, DTI.

Fallows, J. (1984) 'A Parable of Automation', in *New York Review of Books*, 27 September, p. 16.

Finegold, D. and Soskice, D. (1988) 'The Failure of Training in Britain: Analysis and Prescription', *Oxford Review of Economic Policy* 4 (3), 21–53.

Finniston, Sir M. (1980) *Engineering our Future*, Report of the Committee of Inquiry into the Engineering Profession, London: HMSO.

Fores, M. and Rey, L. (1979) 'Technik: The Relevance of a Missing Concept', *Higher Education Review*, Spring, 43–57.

Francis, A. (1988) Seminar on 'Technology and Core-competences', presented at Aston University, October.

Fudge, C. (1986) 'Re-Training for New Technology: Six Success Stories', *Personnel Management*, February.

Gill, C. (1985) *Work, Unemployment and the New Technology*, Cambridge: Polity Press.

Gordon, A. (1985). *Adult Training in Britain*, Manpower Services Commission: Sheffield.

Gunn, T. (1987) *Manufacturing for Competitive Advantage*, Cambridge, Mass.: Ballinger.

Hampden-Turner, C. (1984) *Gentlemen and Tradesmen: The Values of Economic Catastrophe*, London: Gower.

Henderson, J. (1989) *The Globalization of High Technology Production*, London: Routledge.

Hendry, C. (1991) 'Corporate Strategy and Training', in J. Stevens and R. Mackay (eds), *Training and Competitiveness*, London: NEDO/Kogan Page.

Hirst, P. (1989) *After Thatcher*, London: Collins.

Hofstede, G. (1986) 'The Usefulness of the "Organisational Culture" Concept' (editorial), *Journal of Management Studies*, 23 (3).

Hunt, T.L. (1984) 'Robotics, Technology and Employment', in T. Lupton (ed.), *Human Factors in Manufacturing*, Amsterdam: IFS/North Holland.

Jacques, E. (1951) *The Changing Culture of a Factory*, London: Tavistock.

Jenkins, D. and Vandevelde, M. (1985) 'Are The Sights Set Too High?', *Financial Times*, 21 August.

Keegan, V. (1990) 'Barometer of decline that could point to storms ahead for Britain', *Guardian*, 30 May.

Keenan, A. and Newton, T.J. (1986) 'Work Aspirations and Experience of Young Graduate Engineers', *Journal of Management Studies* 23(2).

Keep, E. (1989a). 'A Training Scandal?' in K. Sissons (ed.), *Personnel Management in Britain*, Oxford: Blackwell.

——(1989b) 'The Grass Looked Greener – Some Thoughts on the Influence of Comparative Vocational Education and Training Research on the UK Policy Debate', Paper presented at Training Agency Conference on International Comparisons of Vocational Education and Training for Intermediate Skills, Manchester, September.

Kern, H. and Schumann, M. (1987). 'Limits of the Division of Labour', *Economic and Industrial Democracy* 8 (2), 151–70.

Kolodny H.F. (1985). 'Work Organisation in Sweden', *Human Systems Management* 5 (3), 207–20.

Lane, C. (1988) 'Industrial Change in Europe: the Pursuit of Flexible Specialisation in Britain and West Germany', *Work, Employment and Society*, 2 (2), 141–68.

Lash, S. (1989) 'Postmodernism as a Regime of Signification', *Theory, Culture and Society* 5 (2–3), 311–36.

Lawrence, P. (1980) *Managers and Management in West Germany*, London: Croom Helm.

Leavitt, H.J. and Whistler, T.L. (1958) 'Management in the '80s', *Harvard Business Review* 36, 41–8.

Lindley, R. (1991) 'Individuals, Human Resources and Markets', in J. Stevens and R. Mackay (eds), *Training and Competitiveness*, London: NEDO/Kogan Page.

Littler, C. (1983) 'Conclusions: A History of "New" Technology', in G. Winch, *Information Technology in Manufacturing Processes*, London: Rossendale, p. 135.

Lowe, A. (1975) 'Adam Smith's system of Equilibrium Growth', in A.S. Skinner, and T. Wilson, (eds), *Essays on Adam Smith*, Oxford: Clarendon Press.

McLoughin, I. and Clark, J. (1988) *Technological Change at Work*, Milton Keynes: Open University Press.

Manpower Services Commission (MSC) (1985) *The Impact of New Technology on Skills in Manufacturing and Services*, London: MSC.

Mansfield, R. (1986) *Company Strategy and Organizational Design, London: Croom Helm.*

Marsden, D. and Ryan, P. (1989) 'Initial Training, Labour Market Structure and Public Policy in Britain and the FRG', Paper presented at Training Agency Conference on International Comparisons of Vocational Education and Training for Intermediate Skills, Manchester, September.

Maurice, M., Sorge, A. and Warner, M. (1980) 'Societal Differences in Organizing: A Comparison of France, West Germany and Britain', *Organization Studies* 1 (1), 59–86.

Maynard, G. (1988) *The Economy under Mrs. Thatcher*, London: Blackwell.

Meager, N. (1986) 'Skill Shortages Again in the British Economy', *Industrial Relations Journal* 17 (3), 236–48.

Mumford, L. (1933) *Technics and Civilization*, London: Heinemann.

NEDO/MSC (1984) *Competence and Competition: Training and Education in the FRG, USA and Japan*, London: National Economic Development Office and Manpower Services Commission.

Northcott, J. (1986) *Microelectronics in Industry: Promise and Performance*, London: Policy Studies Institute.

——and Rogers, P. (1984). *Microelectronics in British Industry: The Pattern of Change*, London: Policy Studies Institute.

Okubayashi, K. (1984) *Microelectronics Employment and Work Organisation in Japan*, Working Paper 8406, Kobe University, Japan, pp. 32–3.

Ormerod, P. and Salama, E. (1990) 'The Rise of the British Underclass', *Guardian*, 19 June.

Pearson, R., Andreutti, F. and Holly, S. (1990) *The European Labour Market Review; the Key Indicators*, IMS Report No. 193, University of Sussex: Institute of Manpower Studies.

Peitchinis, S. (1983), *Computer Technology and Employment: Retrospect and Prospect*, London: Macmillan.

Peters, T.J. and Waterman, R.H. (1982) *In Search of Excellence*, New York: Harper & Row.

Piore, M. and Sabel, C. (1985) *The Second Industrial Divide: Possibilities for Prosperity*, New York: Basic Books.

Pollard, S. (1982) *The Wasting of the British Economy*, London: Croom Helm.

Prais, S.J. (1981) 'Vocational Qualifications of the Labour Force in Britain and Germany', *National Institute Economic Review* 98, 47–59.

——and Wagner, K. (1983) 'Some Practical Aspects of Human Capital Investment: Training Standards in Five Occupations in Britain and Germany', *National Institute Economic Review* 105, 46–63.

Rainbird, H. (1990) *Training Matters*, London: Blackwell.

Ray, G. (1986) 'Innovation in the Long Cycle', in R. Roy and D. Wiell (eds), *Product Design and Technical Innovation*, Milton Keynes: Open University Press.

Reve, T. (1990) 'The Firm as a Nexus of Internal and External Contracts'. in M. Aoki, B. Gustaffsson and O. Williamson (eds), *The Firm as a Nexus of Treaties*, London: Sage.

Rose, R. (1990a) *Prospective Evaluation through Comparative Analysis: Youth Training in a Time–Space Perspective*, University of Strathclyde: Centre for the Study of Public Policy.

——(1990b) 'The Need for Master Workers', *Financial Times*, 30 July.

Rothwell, S.G. (1984) 'Company Employment Policies and New Technology' in M. Warner (ed.), *Microprocessors, Manpower and Society*, Aldershot: Gower.

——and Davidson, D. (1983) 'Training for New Technology', in G. Winch (ed.) *Information Technology in Manufacturing Processes*, London: Rossendale.

Sabel, C.G. and Zeitlin, J. (1985) 'Historical Alternatives to Mass Production: Politics, Markets and Technology in Nineteenth Century Industrialization', *Past & Present* 108, August, 133–77.

Schumann, M. (1990) 'Changing Concepts of Work and Qualifications' in M. Warner, W. Wobbe and P. Brödner (eds) *New Technology and Manufacturing Management*, London: Wiley.

Senker, P. (1984) 'Coping with New Technology: The Need for Training', in T. Lupton. (ed.) *Human Factors in Manufacturing*, Amsterdam: IFS/North Holland.

——(1985) 'Some lessons from research', in P. Senker (ed.), *Planning for Microelectronics in the Workplace*, London: Gower.

——(1988) 'International Competition, Technical Change and Training', Papers in Science, Technology and Public Policy No. 17, Science Policy Research Unit, University of Sussex.

——(1989) 'Technical Change, Work Organization and Training: Some Issues Relating to the Role of Market Forces', *New Technology, Work and Employment* 4 (1), 48–55.

——(1990) 'Some Economic and Ideological Aspects of the Reform of Education and Training in England and Wales in the Last Ten Years', *Journal of Education Policy* 5 (2), 113–25.

——and Beesley, M. (1986) 'The Need for Skills in the Factory of the Future', *New Technology, Work and Employment* 1 (1), 9–17.

Shutt, J. and Whittington, R. (1987) 'Fragmentation Strategies and the Rise of Small Units', *Regional Studies* 21 (1), 13–23.

Smith, A. [1776] (1975) *The Wealth of Nations*, London: J.M. Dent.

Smith, C. (1984) 'Design Engineers and the Capitalist Firm', working paper, Work Organisation Research Centre, Aston Business School.

Smith, M. (1990) 'Stark Choices in the Labour Market', *Financial Times*, 3 December.

Sorge, A. (1984) *Technological Change, Employment, Qualifications and Training*, Berlin West: CEDEFOP, p. 23.

——Hartmann, G., Nicholas I. and Warner, M. (1983) *Microelectronics and Manpower in Manufacturing*, Aldershot: Gower.

——and Warner, M. (1986) *Comparative Factory Organisation*, Aldershot: Gower.

Steedman, H. and Wagner, K. (1987) 'A Second Look at Productivity, Machinery and Skills in Britain and Germany', *National Institute Economic Review*, November, 84–95.

Stevens J. and Walsh, T. (1991) 'Training and Competitiveness', in J. Stevens and R. Mackay (eds), *Training and Competitiveness*, London: NEDO/Kogan Page.

Tarbuck, M. (1985) 'The Engineering Industry', in P. Senker (ed.), *Planning for Microelectronics in the Workplace*, London: Gower.

Tarsh, J. (1985) 'Graduate Shortages in Science and Engineering', Department of Employment, Research Paper No. 50.

Thompson, H. and Scalpone, R. (1985) 'Managing Human Resources in the Factory of the Future', *Human Systems Management* (3), 221–30.

Wagner, K. (1983) *Relations Between Education, Employment and Productivity and their Impact on Education and Labour Market Policies: A British–German Comparison*, Berlin: European Centre for the Development of Vocational Training.

Warner, M. (1985) 'Microelectronics, Technical Change and Industrialized Societies: An Overview', *Industrial Relations Journal* 16 (3), 19–33.

——(1986) 'Human Resources Implications of New Technology', *Human Systems Management* 6 (4), 279–88.

Wickens, P. (1991) 'Innovation in Training Creates a Competitive Edge' in J. Stevens and R. MacKay *op. cit.*

Wilkinson, B. (1983) *The Shopfloor Politics of New Technology*, London: Croom Helm.

Williams, K., Williams, J. and Haslam, C. (1986) 'Flexible Specialisation and the Future of British Manufacturing', Paper presented at the Conference on Flexible Specialisation, Birkbeck College, 14–16 July.

Williams, K., Williams, J., Haslam, C. and Wardlow, A. (1989) 'Facing up to Manufacturing Failure'. in P. Hirst and J. Zeitlin (eds), *Reversing Industrial Decline?*, Oxford: Berg.

Williams, V. (1984) 'Employment Implications of New Technology', *Employment Gazette*, May, 201–15.

Williamson, O. (1975) *Markets and Hierarchies*, New York: Free Press.

Winch, G. (1983) 'Introduction' to G. Winch (ed.), *Information Technology in Manufacturing Processes*, London: Rossendale.

Woodward, J. (1965) *Industrial Organisation Theory and Practice*, Oxford: Oxford University Press.

Worthington, B. (1989) 'Educating and Training Engineers', in B. Burnes and B. Weekes (eds), *'AMT: A Strategy for Success?'*, London: NEDO.

Author index

Subject index